Indian Marine Biology

G. SESHAPPA

2026

Daya Publishing House®

A Division of

Astral International Pvt. Ltd.

New Delhi – 110 002

First Published, 1991
Reprinted, 2026

ISBN: 978-81-7035-098-9 (Harbound)

Published by : **Daya Publishing House®**
A Division of
Astral International Pvt. Ltd.
– ISO 9001:2015 Certified Company –
4736/23, Ansari Road, Darya Ganj
New Delhi - 110 002
Phone: 011-4354 9197, 2327 8134
E-mail: info@astralint.com
Website: www.astralint.com

Digitally Printed at **Replika Press Pvt. Ltd.**

This book is dedicated to the memory of late Shri D. H. Veeraiah of the Indian Accounts and Audit Services who was a Senior Accountant General for several years and who was my classmate at College, colleague at the University, & my friend and general sympathiser during the rest of his life.

Acknowledgements

I am grateful to Dr. R. Raghu Prasad, Retired Asst. Director General (Fisheries) of the Indian Council of Agricultural Research for going through the draft of this book and offering his kind suggestions for its improvement and also for giving me a Foreword for it readily when asked. I am highly indebted to Dr. S. Jones, former Director of the Central Marine Fisheries Research Institute who when he was Director, gave me the opportunity of being Head of the Division of Fishery Biology of the entire Institute for several years and also made me at the same time the Officer-in-Charge of the Calicut Substation of the Institute, as these assignments gave me the chance to acquaint myself with the work of all Fishery biology scientists of the Institute and also the work of all divisions of the Calicut Centre. I am thankful to Dr. E.G. Silas, Dr. S.Z. Qasim and Dr. R.V. Nair also for similar encouragement when they were Directors after Dr. Jones.

No work of this kind will be possible without drawing from other published sources. In fact this is only a compilation of what others have said, with my own slant and orientation for the subject. I owe a deep debt of gratitude to all sources thus used and they are all cited in the Bibliography. But I must make a special mention of the works of Lagler et al (1962), Qasim (1970), Silas (1982) and Ramachandran Nair and Pillai (1987) from which sources I have cited profusely. I am also grateful to my wife and to my friends who have given me lots of moral support and encouragement in my effort.

G. SESHAPPA

Foreword

The rich and fascinating variety of plants and animals in our seas, makes them a veritable naturalists' paradise. For the uninitiated, the entire area would appear to be just sand, water and waves. In many foreign countries the art of watching and studying the under-sea animals and plants in their natural habitat with *snorkel* or *scuba*-diving is extensively practised even by amateurs. This is yet to become popular here although under the aegis of the MARINE BIOLOGICAL ASSOCIATION OF INDIA an attempt to form a "Snorkel Club" was made some years ago.

Marine Biological Studies of the seas around India were started several decades ago and Dr. G. Seshappa has given a good historical account of these. These investigations have provided a wealth of information and this has been substantially supplemented by the results of the International Indian Ocean Expedition. But much more remains to be done to gain a fuller understanding of our Mahasagar. These studies, though expensive, are interesting, valuable and rewarding, and it is a matter of gratification that several organisations in our country are currently engaged in this task.

The information available so far is scattered mostly in scientific Journals and not much effort has been made to bring this together in the form of books addressed to the layman as well as the Scientist. It is in this context that this book on "Indian Marine Biology" by Dr. Seshappa assumes importance; it fulfills a longfelt need for a concise general yet fully scientific account of our marine organisms, methods of collecting them, and their industrial and commercial uses. This illustrated book not only covers these organisms, (especially the animals), but also highlights the more important fisheries along the coast of India.

It is hoped that this book will stimulate interest especially among the younger generation, to get a better appreciation of the wealth and beauty hidden in our seas - the nature's endowment.

R. RAGHU PRASAD
Retd. Asst. Director General (Fisheries)
Indian Council of Agricultural Research
New Delhi

Preface

My first acquaintance with the sea was when I was a college student; the teachers had taken out a student party to Pamban near Mandapam Camp for field work and included me also. We camped there for over a week and made collections of animals from the shore. What I learnt was of course very little in this short course but it was a landmark in the life of an inland youngster moving towards the sea! When I got the chance of teaching Zoology (and biology) to Intermediate students for over five years from 1940 onwards, I had occasion to study a little more of the sea creatures (in addition to others), a much clearer understanding of things being found more essential at that stage than as a student whose aim was mainly to get through the examinations. I later got a chance during training in marine biology and fisheries, to work for a doctorate in England; I worked right down on the cold seashore of north-east England for the best part of two and half years, and it was indeed a thrilling experience and education about the creatures of the shore, both plants and animals. On return from England in the last quarter of 1948, I got a job in the Central Marine Fisheries Research Institute of Mandapam to do some marine biological and fish (and fisheries) research work. I continued with the Central Marine Fisheries Research Institute till the age of 60 years and retired in 1978 as a Fishery Scientist which post I had held for over 19 years being at the same time the Officer-in-Charge of the Calicut Research Centre of that Institute.

My own studies at the CMFRI concerned some aspects of the benthos and hydrology in the beginning along with the fishery of the Malabar sole; the latter work was continued after some interruption and on a reduced scale till the end of my service, and along with that I worked on some aspects of the Indian mackerel also at Karwar and Calicut. While at Bombay for some years I worked with the Offshore Fisheries Unit doing studies on the Sciaenid fishes. In addition to my own work, from 1959 onwards my duties involved the co-ordination of the research work of some colleagues in all sections of work and thus I learnt something of all aspects of marine biological and fishery-biological research work and the achievements of the Institute in these lines. In due course prompted by friends, I desired to write a book to be readable by advanced scientists as well as the educated lay persons on marine biology and fisheries, and began making some notes from time to time for the proposed book. Now after over 12 years after retirement from service I am offering this book to readers with the saying: "better late than never"! I was in fact thinking of postponing this further with the hope of making it a more exhaustive compilation of uptodate knowledge of the sea and its biology, but considering my age (73 years) and bad general health I have sent this to the press without further waiting. If I do live to complete a bigger volume of writing of a useful nature, I shall certainly bring it out in due course.

Much knowledge has accumulated about sealife in recent years, especially after our attainment of Independence and after the International Indian Ocean Expedition of the nineteen sixties; but books in this line are still few at present and this one may yet be not redundant.

I feel there is no need to offer comments on the contents of the book here; they will become clear to the readers as they go on with the book which must provide its own justification. It was not written with any University or college standard in view for a text book, I myself being not acquainted with the curricula of the various college and University standards of these days. But if it is found useful for any particular standard such as the PUC or the degree level in a general way it will be very fine indeed; but there are portions in the book which will interest postgraduate students and research scientists also, such as the problem of catching intersurface fauna at the seabottom and the mud bank phenomenon and productivity etc.

There will be many mistakes even in this small book; I shall be thankful to readers who will kindly bring such mistakes to my notice so that I may set them right in the next print or edition.

G. SESHAPPA

Contents

List of Illustrations

Chapter I

Introduction

The sea is an immense body of water; depending on this water and living inside it are many different kinds of plants and animals. All sizes and grades of organisms beginning from the most minute and ultra-microscopic unicellular creatures right upto the largest living animals of the present day namely, the whales and whale sharks are included among these creatures of the sea. It is believed that the first beginnings of living material appeared in the sea itself and that life gradually evolved and invaded the land. In addition to the various kinds of plants and animals that are living at the present time, many fossil remains of extinct creatures of the past ages also occur on the seafloor.

The pattern of life in the sea is clearly of a unique kind and this must naturally be adjusted in accordance with peculiarities of the marine environment. The creatures of the sea are divisible into different categories depending among other factors, on the depth or region of their occurrence and their special modes of living. There are for instance, animals which live attached to substrata at the seabottom or burrow into the sand and mud there, or show only limited wriggling and creeping movements on the floor of the sea; such creatures (along with any bottom plants that may occur) constitute what is known as the **benthos**. There are others which live in the upper layers of the seawater largely lacking the power of independent movement and therefore just drifting and floating about with the currents and tides of the water. These creatures spend their lives partly or entirely in the upper layers and at the surface of the seawater and are together known as the **plankton**. There are again still others which possess powers of locomotion and spend their lives at the surface or bottom or in the different levels of the columns of the seawater, many being capable of living in all these kinds of situations; they are together known as the **nekton**.

Among the plants of the sea the minute planktonic forms are the most abundant; there are however many benthic plants especially in the littoral and sub-littoral areas, and particularly on the rocky beaches and in the shallow areas of the sea. It may be stated that the sea is very helpful to the inhabitant plants and animals of the different environments because of the fact that most of the requirements for living such as food organisms, nutrient salts, vitamins and dissolved oxygen are all available in the immediate surroundings of the organisms themselves. This does not mean that there is a homogeneous distribution of either the organisms or of their life-requirements in the sea. On the other hand, if the ocean mass is considered as a whole it is indeed very heterogeneous in the distribution of the different plants and animals and their communities, and also in its other biological and physico-chemical properties. Different animals and plants have different requirements for their living and optimum development though there are many common categories of these requirements also; the maximum development of the organisms is found in environments most suited to them. Because of geographical and topographical variations in the various environmental factors, different areas of the sea have different types of organisms and communities of organisms living in them.

Most of the taxonomic groups of animals are represented in the sea as on land and in freshwater, with some important exceptions. Inestimable quantities of single-celled creatures of both the plant and animal kingdoms are known to exist in the sea. The many-celled animals range from the simple Coelenterates and sponges to the whales and seals of the mammalian group, through almost the entire range of the Invertebrate as well as the Vertebrate taxonomic groups. The morphology and physiology of all these creatures are so wonderfully adjusted to sealife that they are perfectly at home in the sea even as the land animals are on land, the freshwater animals in freshwater and the birds in the air.

It is too obvious to need any stress that the most important difference between the sea and freshwater is the quality of the water itself as an environment. The seawater is not sweet or soft as in tanks and rivers but is very salty so as to make it repulsive for the non-marine creatures in general. Because of sharp differences between seawater and freshwater it is unusual to find animals and plants of the sea in freshwater and *vice versa*. However, an exception to this is the estuarine environment which can harbour both marine and freshwater organisms to some extent, these forming a specialized community of estuarine plants and animals by themselves, there being limitations to the extent to which they can extend into the salt water of the sea or into the freshwater of the rivers.

Depending on the depth of the sea as well as the various properties of the environment (such as the properties of the seawater itself, the occurrence of other organisms, nature of the substratum at the seafloor and other oceanographic features) different kinds and sizes of organisms of varying body structures may be found in different areas of the sea. The organisms have to readjust the tempo of their lives or their *biology* as we may call the sum total of their life activities, when they move from seawater to freshwater or from freshwater to seawater; similarly, adjustments of various degrees will have to be achieved by the organisms when they move from one region of the sea to another of markedly different environmental properties. The differences between one area and another in the sea may relate mainly to any or all of the following:- depth of the water, turbulence or calmness of the water surface, turbidity, concentration of various nutrient salts and other dissolved and suspended substances, occurrence or otherwise of various toxic substances such as hydrogen sulphide, occurrence or otherwise of various useful trace elements, the hydrogen-ion-concentration or

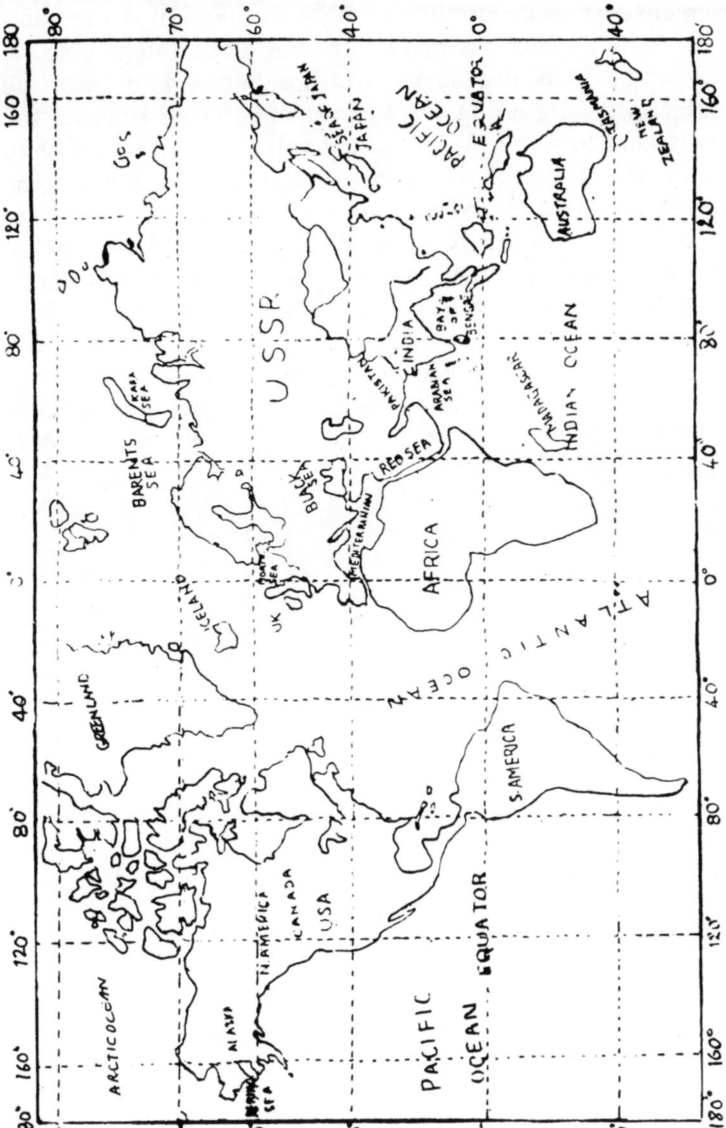

Fig. 1. Map of Oceans and Continents (Semidiagrammatic)

p^H of the water, the intensity and penetration of light at different depths and areas, pressure of the water which depends on the depth, temperature, occurrence or otherwise of food organisms and predators, and also the movements of the water in the form of tides and currents. Depending on the interaction of these various factors and their resultant suitability or otherwise to the particular form or forms of the organisms concerned, it is usual to find some kinds of animals and plants more abundantly in some parts of the sea than elsewhere. All organisms thrive well and abundantly only in the best of conditions required for them; but these conditions are not the same in all places even in limited areas of the sea, but they show variations in quality as well as quantity with change of place and time. Marine Biology is the discipline of study aimed at the understanding of these various environmental factors both individually and in combinations in association with the various organisms of the sea and their life histories which may be influenced at every stage by these environmental phenomena.

Chapter II

History of Marine Biology in India

It is said that even before 1000 B.C. the Phoenicians who were among the first mariners of the world, visited the Indian seas also and attempted to gather valuable information about some parts of our seas. The first sea voyage for modern marine research was made by Captain James Cook during the years 1868-76. This explorer travelled through the seas of the world and found out the delimiting boundaries for the major oceans; he was also the first (European) to cross the Antarctic Circle; he also proved to be first to measure temperatures of the bottom layers of seawater, and in taking measurements of currents, winds and depths of the seas. The first book on Oceanography was published by one Mr. Mathew Maury of America in 1855 under the title: "Physical Geography of the Sea." The foundations of oceanographic research was laid by the "Challenger Expedition" during the years 1872 to 1876. The *Challenger* explored all the seas of the world except Arctic and collected information about the depths of the various regions, the chemical substances found in the water at different places (as could be analyzed in those days) and the benthic animals as well as the nature of the seabottom.

The beginnings of oceanographic studies in the Indian area were also made in the second half of the 19th century. Some English members of the Royal Asiatic Society of Bengal who were engaged in Zoological researches in this country took the initiative in this direction, and as a result of their efforts, an expert committee was appointed by the Government of India in the year 1871 A.D. for going into the question of conducting researches on the deep sea organisms of the Indian seas. The Government started the Marine Survey Department in the year 1874; Owing to insufficient facilities however, this department could not undertake any detailed work until 1881, but the scientists were still able to collect some samples of animals from the sea in depths down to 100 fathoms by means of dredges. In 1872 Wood-Mason was able to collect deep sea animals from the Andaman and Nicobars area on behalf of the Indian Museum of Calcutta. The Marine Survey Department was able to collect samples of about 24 per year down to about 2000 fathoms between the years 1884 and 1894 A.D. using the Research Vessel "Investigator" for the purpose. Samples were collected during this period (by this vessel) of water, animals and also other materials from the Persian gulf, Ceylon, Andamans, Burma and other regions of our seas. The results of most of these researches are described in the publications of the Indian Museum brought out from the year 1898 onwards periodically. An Officer designated the Surgeon - Naturalist was included among the staff of the "Investigator" and much oceanographic and biological work was done by the incumbents of this post, which however was not occupied continuously for presumably administrative reasons. Among the Surgeon-Naturalists that took keen interest in Marine Zoological researches must be mentioned the name of Dr. (Col) Seymour Sewell as well known. Col. Sewell was a world-famous researcher who gave much encouragement to marine studies in our country in the pre-Independence era. He himself led the "John Murray" Expedition on the Research Vessel *"Mabahiss"* for conducting researches on biological and oceanographical problems in the Arabian Sea. The manganese and other metallic nodules which are proving to be a valuable resource from the sea bottom at present, were already discovered in some parts of the Arabian Sea even in the above Expedition, though only still in patches perhaps because of the inexhaustive survey that could not concentrate on any particular aspect excluding others. Col. Seymour Sewell's work on the copepods (a group of animals belonging to the Arthropoda and that are very useful as food of many kinds of fishes in the seas) is even to-day an authoritative reference book on these animals in this area of the world.

From the beginning of this century the fishery scientists of the Government of Madras (which is now Tamil Nadu) and the specialists attached to the Connemara Museum in Madras also conducted some researches on the animals and plants of the sea, especially of the shore. Thurston and Herdman conducted researches independently in the Gulf of Mannar on the south-east coast of India; Gravely (1930, 1941 & 1942) examined the shells and other animal remains from the beach; the littoral fauna of Krusadai Island in the Gulf of Mannar were worked out by various persons including Gravely, Fauvel, Burton and others; these results are detailed in Vol. I, No. 1 of the Madras Government Museum Bulletins (1927 to 1937). P. Fauvel's contributions (1932 & reprint 1953) to our polychaete fauna are very important also. Among other works from the Madras Museum must be specially mentioned those of Burton (1937) on the sponges, Hickson on the Alcyonaria (1931), M.G.K.Menon on the Scyphomedusae (1930 and 1936), Krishna Menon on the Decapod larvae and life histories (1933, 1937 and 1940) and of C.C. John on the Chaetognaths of Madras (1933). Another expert namely, J.S. Gardiner (1903) studied the animals of the Lakshadweep group of islands and published his results in book form. At the beginning of this century, a Fisheries departmental Research establishment was opened at Calicut on the west coast for initiating marine work, and some papers were published by the expert scientists of this research Unit on the fisheries and marine biology of the south-west coast of India. Research Units were also set up in due course at Krusadai near Mandapam and at Ennur near Madras to carry out some limited investigations; these units were also helpful in supplying marine specimens to workers all over the country for routine study as well as research.

Various fishing voyages in the sea have also helped in the progress of marine biological researches in India in some way or the other from time to time. From the year 1928 onwards, the Madras Fisheries Department conducted several exploratory surveys on both the east and west coasts for spotting regions of good fish stocks. In 1928-30 the *Lady Goschen* did the first extensive exploration of this kind on both the coasts. Similarly but even earlier (Hefford 1922) the Bombay fisheries Department conducted exploratory trawling surveys off Bombay and the north west coast with the vessel *William Carrick*. The Government of Bengal had also surveyed the deep sea fishing potentialities using the vessel *Golden Crown* even much earlier (1908-1911). From the experience of these surveys, not only were considerable data made available of the fishery wealth in the respective regions of the sea around India at that time, but some knowledge also became available about the oceanographic conditions and general marine biology in the areas investigated, though the trips of all these vessels were essentially fishing voyages only. In recent years exploratory and fishing voyages of large vessels in the offshore regions of our seas are very common, both Governmental organizations and private business interests being represented among these. Research workers frequently get facilities to accompany the fishing crew in such voyages not only to observe the fishing activity but also to collect samples and data of different kinds for study.

There has been an all-round increase in the facilities for the study of marine biology in our country after the attainment of Independence. The Government of India has started several institutions either directly under its Ministries or as autonomous bodies with Government grants, such as the Central Marine Fisheries Research Institute (which is now a part of the Indian Council of Agricultural Research), the Deep Sea Fishing Station, the Marine Survey of India and the National Institute of Oceanography, each of these having a country-wide distribution of establishment and jurisdiction for work, and contributing directly to the growth of Indian Marine Science.

In addition to the Central Institutions, the maritime States of the Country also have some research activities administered through their fishery departments, the problems of investigation being mainly of local industrial application; the more academic and comprehensive problems of an all-India nature are tackled by the Central Research Institutions. The various Universities in the maritime States of the country i.e. situated in the States of Gujarat, Maharashtra, Goa, Karnataka, Kerala, Andhra Pradesh, Tamil Nadu and Orissa as well as West Bengal, have courses in aquatic fisheries (or purely marine fisheries in some cases) and biology thus providing the staff to man the research Institutes as well as the administrative departments in Fisheries.

A very extensive international marine research expedition was conducted in the sixties in the Indian Ocean, this being unprecedented in many respects, and several countries outside our region also participating in the work. About forty research ships from twenty countries were involved in this gigantic effort at the scientific surveying of the Indian Ocean; regions of the ocean never visited before were also covered and samples and data were collected of the organisms

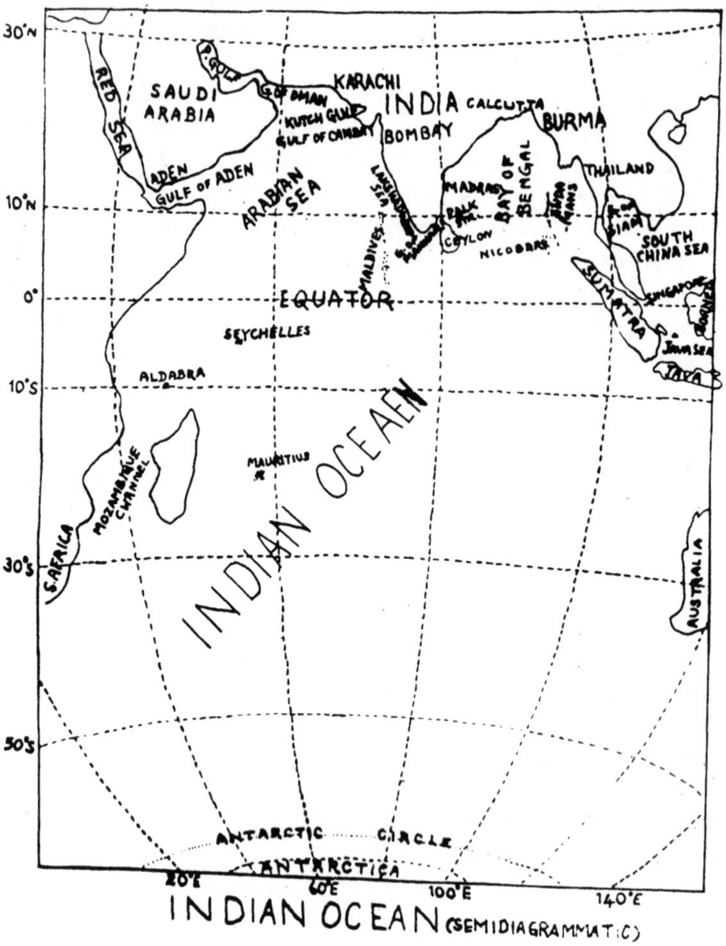

Fig. 2. Map of Indian Ocean (Semidiagrammatic)

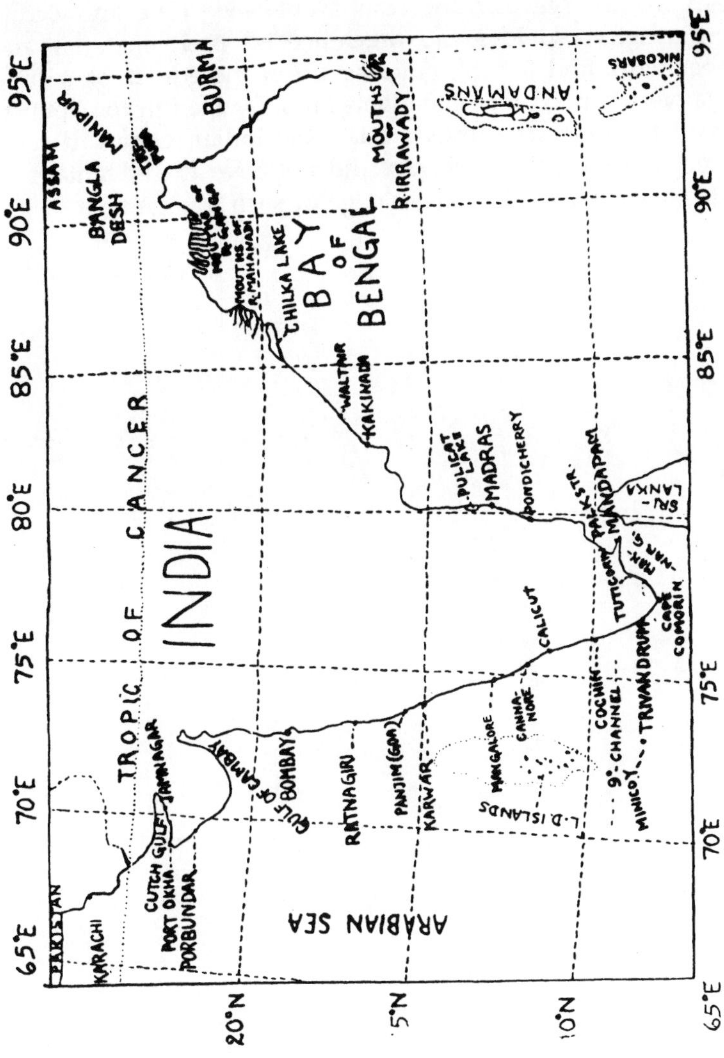

Fig. 3 Map of the Sea and the Coasts Around India (Semidiagrammatic)

of the seabottom, plankton, seawater, mudsamples and various other materials occurring in the different regions of the great ocean. Most of the samples and the data have since been worked out and the results published. The knowledge available from these results provides the much-needed basic material and foundation for further work and for the building up of the edifice of marine biology of our area on very sound and dependable lines, for the future scientists of our country and of the other countries around the Indian Ocean. It need hardly be emphasized that the earlier workers did not have the advantage of such a comprehensive collected fund of knowledge of such a large area of our seas.

Chapter III

Chief Features of the Sea Around India

(*a*) *General and Climatic Features:* The sea around India is part of the great Indian Ocean and it will be seen from the map that the watermasses of the east and west coasts are quite distinct from each other but united together in the south and south-east of the country between itself and Sri Lanka. The Palk Straits and the Gulf of Mannar are situated as parts of this connecting water link between the Bay of Bengal and the Arabian Sea. The latter is the watermass to the west of the country while the Bay of Bengal is the watermass to the east and south-east. As the landmass of India tapers down to the south, these two areas of sea form the southwestern, southern and southeastern boundaries of the country, meeting as stated above, through the Gulf of Mannar and the Palk Strait.

The Arabian sea extends from about 6°E latitude to about 25°N latitude, and from about 51°E longitude to 77°E longitude in the west-to-east direction. Leaving alone the southernmost part, the rest of the Arabian Sea is surrounded by land; at its northwest are the coasts of Arabia and other African countries; in the north is the coast of Pakistan; to the north-east of the Arabian sea are the Gulfs of Kutch and Cambay while to the interior of the Gulf of Cambay is a vast expanse of marshy land known as the Rann of Kutch. As in the case of Bay of Bengal on the east coast several rivers flow into the Arabian Sea on the west coast also, there being said to be as many as 41 rivers (medium and small) on the Kerala coast itself; however some of the rivers on the east coast are very large ones and total volume of freshwater brought into the sea by the rivers is known to be more on the east coast than on the west coast.

About a hundred and fifty miles away from the main land off the Kerala coast on the southwest of the country lying between 8°N and 12°30' N latitude, and 71°E and 74°E longitude is a group of 27 small islands of variable size (17 of these being uninhabited islets only), known as the Lakshadweep Islands; these were formerly known as the Laccadive Islands and have a total area of only 28.5 sq. km altogether (Jones 1986); some of these are larger than the others and ten of them are now inhabited by people also. Minicoy is one of these larger islands and is situated far south in the chain, somewhat separated and distant from the remaining islands of which the main ones may be mentioned as Androth, Kiltan, Kadmat, and Amindivi. It was customary till recently in official correspondence to refer to these Lakshadweep Islands as the Laccadive, Minicoy and Amindivi Islands for administrative purposes, the administrative Offices being housed perhaps from the beginning in the last two islands out of the lot. There is a Unit of the Central Marine Fisheries Research Institute in Minicoy and the Kerala Government has some staff in one or more of the other islands as well as Minicoy. Frequent Steamer services seem to have been started in recent years to these Islands and improvised airstrips are also available for special air services when necessary; all the islanders are said to speak Malayalam, the language of Kerala State, except in Minicoy where the majority of people are said to speak an ancient language called "Mahl" or "Divehi".

The sea near the Lakshadweep islands reaches great depths quite close to the land this being in contrast to the slow and gradual descent of the level of the seabottom from the shore on the mainland of India, the Lakshadweep Islands consisting of coral formations rising steeply out from the sea. Far south of the Lakshadweep Islands and situated as if marking the southern limits of the Arabian sea with the Indian Ocean proper is another group of islands known as the Maldives, which lie quite far away from our country.

The most important climatological aspect of the west coast of India is the occurrence of heavy south-west monsoons which mean very heavy rains often accompanied by strong winds. These rains occur usually from early June to the end of September, July and August being the months of heaviest rain during years of normal rainfall. These rains influence the inshore sea very greatly by causing great surface turbulence with strong waves and currents whose directions depend on the directions of the winds blowing on the surface of the water. The winds and consequently the currents change their directions during different seasons. The wind and rainfall of the southwest monsoon season start from the southwest of the country and move up towards the north and northeast of the seas and of the country gradually along the coast.

The sea around India is also characterized by certain special currents of slowly moving water masses and the phenomenon of water from deeper layers rising upwards under certain circumstances during certain periods of the year; this phenomenon is known as upwelling occurring along the west coast as stray occurrence with maximum between Calicut in the south and Karwar in the north. There is a general clockwise circulation in the northern Indian Ocean under the influence of the landmasses while in the southern Indian Ocean the circulation of water is generally anti-clockwise. During the northwest monsoons there is a well developed north equatorial current while in the southern region there is an equatorial counter current flowing eastward. In summer months the north equatorial current is replaced by the monsoon current flowing from west to east. But eddies developing between these two current systems in a counter clockwise direction in the north of the equator and in a clockwise direction south of the equator.

During the southwest monsoon, the circulation pattern in the Arabian sea is clockwise and a strong southward drift is found along the west coast (especially in the region of Calicut to Karwar). By the end of the southwest monsoon this flow is weakened and a northward current associated with sinking is found along the coast by December. In the Bay of Bengal along the east coast of India, a north-easterly current flows during the period of January to July, and a south-westerly current flows during the period August to December (Jones and Banerji 1973).

The sea between the mainland of Kerala and the Lakshadweep Islands is called the Lakshadweep Sea (or the Laccadive Sea in the earlier days, before the renaming) and is known for its great productivity in terms of fishery wealth. The shelf area estimated for these islands is 4336 sq. km. The seabottom, especially in the shallow areas nearer land, has some very distinctive features, being made of fine soft clay, there being only a few patches of rock and sand in the whole area. In the Kerala inshore waters the soft muddy bottom of the sea often gets agitated or stirred up due to various happenings and this makes the water very turbid; in some places concentrations of mud called the "mud banks" may also be formed during such periods in some places; these mud banks remain for some days or weeks and then disappear, sometimes suddenly and sometimes gradually by the repeated action of the waves and currents and other forces operating in the sea. This unique occurrence in the west coast environment influences the biological cycle in the sea to a very large extent.

The watermass to the east and southeast of the Indian subcontinent is the Bay of Bengal which is bounded on the eastern side by Burma and East Pakistan and on its north and west by the coasts of West Bengal, Orissa, Andhra Pradesh, and Tamil Nadu. The Andaman and Nicobar group of Islands are also part of Indian territory, being of strategic value in the defences of the country. The Andamans are a chain of islands spreading out from north to south, the group being divisible into the north, middle, south and Little Andamans (the last being the southernmost in the chain). Some of the islands are inhabited in the groups of these islands while some are said to be still uninhabited. Port Blair which is the administrative capital of the Andamans lies in the southern Andamans and is a modern city like many on our mainland, with sea and air passenger services regularly to the mainland of India through West Bengal and Tamilnadu.

The Nicobar group of islands are situated further south of the Andamans, to the northwest of the Sumatra islands and to the west of the territories of Malaya and Thailand. Fishing has been in vogue in the Andamans as well as the Nicobars sea areas from times immemorial and some foreign countries are said to have exploited the fishery resources of these sea areas in the past. A Substation of the Central Marine Fisheries Research Institute of the ICAR (under the Government of India) has been working from Port Blair from the early fifties onwards with expanded activities in recent years. The main objective of the research establishments in the Lakshadweep and Andaman groups of islands is the collection of fishery and marine biological data for comparative study with the data that are being regularly collected at the mainland stations. Local problems of a short term nature are also often tackled according to necessity and facilities available from time to time.

The sea around the island of Sri Lanka, Andamans and the Nicobars are all considerably to the north of the equator but still to the south of the tropic of cancer only (unlike the northernmost part of the Arabian Sea) and have mostly tropical climatic conditions. One peculiarity noticed in the Bay of Bengal is the frequent occurrence almost regularly every year, especially during the season of the northeast monsoons, of very severe storms (sometimes accompanied by high waves and tidal bores also). Instances of human mortality and loss of property caused by such storms are also quite frequent during some years.

(*b*) *Different parts of the sea and the conditions of life in the same*: Depending on the depth of the water as we proceed from the shore outward into the sea, the chief regions of the sea are the littoral (intertidal or shore) area, the continental shelf and the continental slope and the deep sea. The littoral region is the margin of the sea or actually the patch of land over which the tides rise and fall regularly at known intervals. The sublittoral region comes between the shore and the continental shelf and extends from the lower limit of the shore to a variable distance out into the sea. Beyond the sublittoral region to a depth of about 200 metres the seabottom slopes gradually and this region is the continental shelf. Beyond this shelf the slope of the seabottom is steep and leads on to the deep sea zone. The extent of the continental shelf is not uniform in all places but varies a great deal from region to region; the continental slope may commence only a short distance from the shore as around the atoll formations of the Lakshadweep islands, or it may commence only beyond many miles (upto even a hundred or more miles) away from the shore. Along the Indian coast the shelf area is considerably less than what it is along the coasts of many other countries, and is also much less on east coast than on the west coast. The total area of the continental shelf has been estimated as 338260 sq. km. in the Arabian Sea and Lakshadweep Sea, and 281987 sq. km. in Bay of Bengal (Jones, 1973). Fishery wealth is known to be most abundant and commercially best exploited in all countries in the continental shelf only, including also the near shore and shallow water areas. Various animals and plants seem to thrive best only inward of the deep sea that is, in the continental shelf region, taking both the seabottom and the column water into account. It appears that the conditions for the existence of animals and plants are optimum only *upto the commencement* of the abyssal zone *but not in* the abyssal area (or zone). One dominant feature of the environment at this depth is the complete absence of sunlight in the lower layers of the water; The very high pressure of upto 1500 atmospheres per square inch of water is the other most important feature of the water at the seabottom.

Many variations are possible in the modes of living of organisms depending upon minute differences in the environmental and other factors. On the seashore itself different kinds of animals and plants thrive in different situations depending on whether the substratum is rock, mud, sand or different grades and combinations of these different types. The larger intertidal plants are mostly abundant mainly or only on the rocky shores. The shore areas around India abound in variations of different kinds of substrata though the rocky shores seem to be considerably restricted. In the shallow seas not too far away from the shore, the sun's light penetrates abundantly into all depths of the water; there is no scarcity of sunlight along our coasts all of which enjoy a tropical climate (except the northernmost part of the Arabian Sea (which is semi-temperate in its climatic features). In the temperate and cold countries of the world as a rule, there is not enough sunlight penetrating the different depths of the sea even in relatively shallow waters, the biology of such areas being therefore considerably affected. Coming to other life-requirements, it must be stated that in the shallow regions of our seas, all environmental factors including the temperature of the water, concentrations of various nutrient and other salts, concentration of dissolved oxygen and other gases, density of food items including both animate and inanimate objects, extent of water currents and wave force, and the variations in quantity of food organisms resulting from the variations and combinations of these factors from time to time, will all be having their influence in the visible total picture at any given time as well as over a period of time. Experience so far has shown that the shallower sublittoral area of the continental shelf is the actual zone of abundance of fish and other organisms so far in our seas. It is in this sublittoral region that there are heavy concentrations of the waste materials and nutrients washed in from the rivers along the different parts of the coast. These materials do not go unutilized in the chemistry and life-cycle in the sea, but do actually turn into very useful and rich fertilizers helping in the increase and maintenance of the animal and plant stocks. All our important seafishes which are of commercial importance are at present caught in some abundance in the shallow areas just mentioned.

In the oceanic regions there can be great differences in environmental conditions as well as in the concentration and productivity of different kinds of organisms, the phytoplankton falling down suddenly in quantity as the light decreases, and followed by the decrease in the zooplankton which is dependent on the phytoplankton for its food. Certain properties such as the salinity and temperature may show little or no change beyond some depths of the water, but pressure increases steadily with depth, this also being one of the important factors affecting life at such depths. Several organisms among both phytoplanktonic and zooplanktonic elements are known to adjust themselves to changes in light intensities during different periods of the day by migrating to different depths of water according to necessity.

Chapter IV

Some Common Physical and Chemical Properties of Seawater

(a) Physical properties

Temperature, density, viscosity, turbidity, light and pressure are some of the commonest physical factors that may affect the life of plants and animals in the sea. Compared to land, the fluctuations in the temperatures are less in the sea because of the ability of the seawater to retain heat longer than land, this difference being true whether the problem is considered on a daily, seasonal or a latitudinal basis. The lowest temperature of sea is -1.9°C and is found in the Antarctic. In the equatorial water of the Red Sea, Arabian Sea, Persian Gulf and other areas, the temperature is high and may reach 35°C; in some isolated lagoons of the Red Sea the summer temperature may even reach 40°C. The temperature decreases with the increasing depth of the water in the sea—except where there are special features such as upwellings and currents which may change their character—, but in the deep sea beyond 200 metres depth the temperature of the water may be said to be relatively stable, only small differences being noticed with very large differences in the depth.

The density of seawater is slightly higher than that of freshwater because of its salt content. In inshore waters the density may show some variation but the normal range of specific gravity may be said to lie between 1.00 and 1.21 only. The turbidity is also variable from time to time in the inshore waters depending on the state of the tide, the winds and several other factors. The nature of the seabottom may also affect the transparency; the degree of turbidity and transparency may be helpful or harmful within limits depending on particular circumstances in each situation. The pressure of seawater increases with the depth of the sea while light decreases with depth. In our inshore sea areas however, it may be said in general that there is abundant sunlight available for the marine organisms.

(b) Chemical composition of seawater

More than most other factors, the chemical composition of seawater influences the life of the creatures of the sea in the various stages of their life histories. Various salts and other substances are contained in the seawater, some

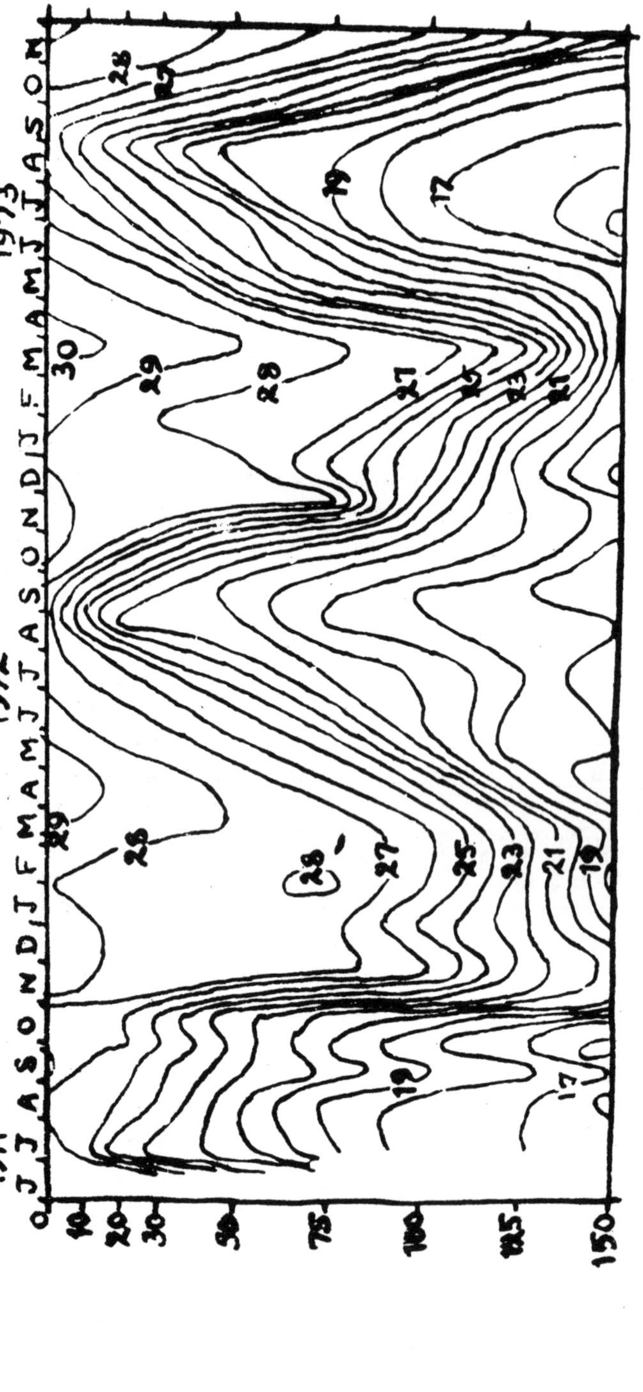

Fig. 4. Mean Temperature Distribution in the Shelf Waters off Cochin, June 1971 to November 1973 (with grateful acknowledgements to UNDP/FAO Pelagic Fishery Project Cochin, from whose *Progress Report* No. 6— "Survey Results 1972-73" the figure is re-drawn)

in dissolved form and others as particulate matter in fine suspension, these including both organic and inorganic materials. The average percentage of total salts in seawater has been estimated as 3.4325; the chlorinity or chlorine content forms 1.90% in the seawater. The values of some of the more important among other constituents in the seawater are follows:

Table 1
Some Constituents of Seawater

Chemical (Ions)	*% in total salts*	*Concentration (g/kg of water)*	*Concentration at 20°C(g/l of seawater)*
Sodium	30.61	10.556	10.809
Potassium	1.10	0.380	0.389
Magnesium	3.69	1.272	1.303
Calcium	1.16	0.400	0.410
Strontium	0.04	0.0085	0.013
Borate	0.07	0.026	0.027
Chlorine	55.04	18.980	19.435
Sulphate	7.68	2.649	2.713
Carbonate	0.41	0.140	0.143
	(Variable depending on hydrogen ions)		
Bromine	0.19	0.065	0.067
Fluorine	0.004	0.001	0.001

(The values are taken from J.A. Colin Nicols (1967): *The Biology of Marine Animals*)

There is some degree of inter-relationship among many of the common properties and constituents of seawater so that, in some cases when the main constituents of a chain of properties are known the remaining one or more properties can be roughly estimated or guessed so as to be enough for many practical purposes. However, in view of variation of these relationships from station to station, it is a normal practice to determine each property separately only. Thus though the density, salinity and temperature of the water are considerably inter-related, they are all separately determined in routine practice.

The salinity or total salt content of seawater is estimated at present by determining the chlorinity content by the titration method. The relative constancy of the constituents of the water in different areas of the sea or within particular masses of the seawater is due to the mixing up of the water by circulation or other means of physical and chemical stabilizing phenomena. Changes in the composition of the water are more common in the inshore or littoral areas than in the oceanic and abyssal regions of the sea.

In the surface waters of the oceanic regions, salinity has been known to show variation generally between 3.3 and 3.17 per hundred parts; the average salinity of the sea may be said to be 35 parts per thousand of the water in the sea. But these values may also be exceeded in special situations at either end. The salinity may be very low where there is heavy precipitation (i.e. rainfall or snow) and in situations where rivers empty their freshwater into the sea. On the west coast of India during the months of the south west monsoon i.e. June to September and especially in July and August, the salinity and temperature of the seawater shows a steep drop in values from the shore outwards down to depth of several fathoms in the sea. As the monsoon rains occur regularly during certain months every year, the lowering phases of the salinity are also correspondingly regular, and there are occasions when the salinity goes down to 10 parts or even less at times per thousand (as noticed at Calicut on the Kerala coastline on several occasions). The sea at Calicut is subject to the influence of two rivers emptying themselves into it with an interval of only a few miles of coastline between the two rivermouths that open out into the sea. These are fairly large rivers compared to many of the total of 41 rivers joining the sea along the Kerala coast, and are at Elathur about 4 miles north of Calicut and at Beypore, about 8 miles to the south of Calicut; they can perhaps be classified as medium (small) sized rivers on an all-India scale. The diluting effect of these two rivers is in addition to precipitation as rainfall which averages to around 120 inches per year, more than half of this total falling within the two or three months of the peak southwest monsoon season. Under such circumstances almost freshwater occurs near the shore for a short duration; in response to this adverse happening many of the animals of the shallow sea near the shore (and upto the depth of 8-10 fathoms) migrate into the more offshore areas where the salinity and temperature are more congenial to them. Animals that are unable to move away as above to areas of normal salinity and temperature conditions may have to face great odds for their very survival and instances of death of the more delicate-bodied creatures are also quite common (Seshappa, 1953).

The Red Sea is an example of an area of very high salinity which goes upto 4.1%; in recent years explorers have discovered deep sea pockets in this region where the salinities may go even higher up. High salinities are as dangerous as low salinities in general, though the physiological effects of the these on the organisms may naturally be in the opposite directions. Theoretically it is not impossible however, for the organisms to adapt themselves in due course to both kinds of situations provided a stability is achieved in either one or the other direction and retained as such for a sufficiently long time, such a state being not impossible in the deeper waters of the ocean.

In addition to the chemical ions listed in table 1 above, there are several other items occurring in relatively small quantities under normal conditions of the sea. Some of these are extremely important for the organisms even as traces. It may be added also that among the other contents of the seawater are included the precious manganese nodules which are at present being commercially exploited, and even precious metals like silver and gold are said to occur in the sea though these are known at present in very minute quantities only.

Many workers have studied the more important physico-chemical properties of the sea around India from various centres, this being particularly so for the waters of the inshore or shallow water fishing grounds around the coast. Annigeri (1968) found at Karwar that the salinity of the inshore waters taken at three depths (surface, 5 fathoms and 6.5 fathoms) in 1964 were in the range of 1.17$^{\circ}/_{o}$ - 3.65$^{\circ}/_{o}$, 1.79$^{\circ}/_{o}$ - 3.66$^{\circ}/_{o}$ and 2.25$^{\circ}/_{o}$ - 3.72$^{\circ}/_{o}$ respectively; in 1965-66 these values were: 2.12$^{\circ}/_{o}$-3.54$^{\circ}/_{o}$, 0.57$^{\circ}/_{o}$-3.54$^{\circ}/_{o}$ and 0.39$^{\circ}/_{o}$-3.60$^{\circ}/_{o}$ respectively. The highest salinity for the inshore waters of our seas is found in the summer months while the lowest is found in the months of heavy monsoon rains (the southwest monsoon on the west coast and the northeast monsoon on the east coast). The monthly trends of temperatures can vary within the seasons from year to year depending on the rainfall trends themselves.

Table 2 shows the monthly mean values of the salinity, dissolved oxygen and phosphates in the inshore waters of the Gulf Mannar on the east coast of Tamilnadu during 1950-52.

Table 2
Salinity $^{\circ}/_{oo}$, Oxygen ml/L and Phosphate ug.at./L during 1950-52 in the inshore waters of Gulf of Mannar

Months	*Surface waters*			*Bottom waters*		
	Salinity	*Oxygen*	*Phosphate*	*Salinity*	*Oxygen*	*Phosphate*
January:	29.04	3.42	0.20	30.22	4.27	0.23
February:	30.35	3.98	0.23	31.43	3.86	0.26
March:	31.58	3.90	0.24	31.82	3.84	0.22
April:	33.77	3.77	0.17	33.27	3.72	0.20
May:	35.54	3.74	0.18	35.17	3.83	0.19
June:	35.40	3.86	0.26	35.41	3.85	0.26
July:	35.45	3.72	0.23	35.29	3.72	0.23
August:	36.18	3.91	0.23	36.15	3.95	0.20
September:	36.10	3.81	0.22	36.00	3.73	0.20
October:	36.11	3.62	0.18	36.11	3.75	0.17
November:	32.83	3.92	0.17	33.01	3.82	0.20
December:	28.88	3.83	0.15	29.40	3.78	0.15

Monthly mean temperatures as given by Prasad (1959) for the Gulf of Mannar showed a maximum of 31.50°C and a minimum of 24.00°C in 1950 while these values were 30.60°C and 24.60° respectively in 1951. The highest temperatures were noticed in April-May while the lowest temperatures were seen during December-February.

The normal range of variation of temperatures in the inshore waters of the Indian sea on both the coasts may be said to be roughly from 25°C to 31°C, specially lower values being noticed in the months of heavy rains during the southwest monsoon particularly along the Karnataka and Kerala coasts.

The dissolved oxygen content of seawater may normally be about one to two ml/L or even less at times but the range may be anywhere from 0 to 5 or more ml/L, very high values being noticed in surf waters of the shore. the 0-values are rather rare and are encountered in regions of high poisonous gases like hydrogen sulphide produced in polluted areas of the sea; along the Kerala coast large quantities of coconut husks are immersed in shallow water near the shore in many places for the fibre to become easily removable; in such areas often a foul smell is noticed which is a sign of pollution. Some kinds of factory effluents may also cause a high reduction in the oxygen content of the waters with a consequent degradation of the biological productivity of the area. High mortality of the living organisms is sometimes noticed in such areas including fishes and other useful animals as well as plants.

The distribution of oxygen does not always seem to follow any clear seasonal trends as may be seen from Table 3. The table taken from Rao (1967) shows the monthly mean values of the temperature, salinity and oxygen in two square areas of the Arabian Sea (squares 9°N - 75°E and 9°N-76°E) for the period 1957-63 and these can be taken as roughly representative of the offshore waters of the southwest coast of India. Fig. 7 shows the average monthly salinities at West Hill, Calicut during the period December 1949 to November 1950.

Phosphates occur in seawater in various forms such as organic, inorganic, dissolved and adsorbed forms and also the particulate form; while these different fractions have their importance individually, the total phosphates give an idea of the complete quantity of the phosphate ions available in any given body of water and this is therefore given more importance by some investigators. Seshappa and Jayaraman (1957) found that total P-values on the west coast at Calicut varied from 2.1 ug-at.-/L to 12.8 ug-at./L while on the east coast the variation was in the range of 1.7 ug.-at./L to 3.0 ug.-at. /L (both determinations being for the years 1950-52). It does not of course mean that the same mean values will be found every year at the respective places, there being variations depending on the biological and other conditions in the areas under observation. Studies seem to indicate that there is a higher uniformity (though the values are much lower) in the phosphate content of the east coast waters than the west coast waters. But it has been found also that the overall values of phosphates found in the sea along both our coasts are many times higher than those found in some other parts of the world. It has also been found that variations in the abundance in the production of phytoplankton in the sea is related among other factors to the fluctuations in the total phosphate values at different times, this latter being largely a measure of the subsequent secondary and tertiary productivity of the sea region concerned.

Table 3

Mean values of temperature (°C), salinity (°/$_{oo}$) and dissolved oxygen (ml/L) in squares 9°N-75°E and 9°N-76° E in 1957-63

Months	*Square 9°N - 75°E*			*Square 9°N - 76° E*		
	Temperature	Salinity	Oxygen	Temperature	Salinity	Oxygen
1	2	3	4	5	6	7
January:	28.29	33.22	4.50	28.21	33.16	4.43
February:	28.42	33.71	4.81	28.71	33.75	4.34

(Contd.)

1	2	3	4	5	6	7
March:	29.63	34.28	4.60	29.69	34.56	4.48
April:	30.40	34.17	4.26	30.66	34.48	3.95
May:	30.44	34.51	4.20	29.83	34.55	3.61
June:	29.30	34.98	4.10	28.73	34.19	—
July:	—	—	—	27.00	23.73	—
August:	26.26	33.38	4.60	25.64	31.56	4.69
September:	26.95	34.43	4.97	26.70	33.19	5.06
October:	27.03	33.50	4.43	27.72	31.39	4.77
November:	28.67	34.24	4.52	28.68	34.42	4.83
December:	28.46	33.86	4.40	27.53	33.49	4.54

The hydrogen-ion-concentration of seawater expressed as p^H (=Log 1/H) a function of the hydrogen ions present in the water is determined as a routine for all water samples in the sea as this value indicates the acidity or alkalinity of the water. Freshwaters have p^H values below 7 while seawater is alkaline and has p^H much higher upto usually around 8 but reaching even 8.7 at times.

Chapter V

The Organisms of the Shore and their Mode of Life

(a) *Tides*: The seashore is a unique type of environment not found in any other situation in the entire world. In this place where the land and sea meet, the water is seen to be rising and falling alternately and in most places twice a day regularly; these ebbs and flows of the water at the margin of the sea are known as the *tides* and are the main contributory factors for the special peculiarity of this environment. The range or amplitude of the tides is maximum during the fullmoon and newmoon days and lowest during the middle periods between these two phases of the moon; the tides of the highest amplitudes are known as the *spring tides* and those of the lowest amplitudes are the *neap tides*; between these two phases of the tides there is a gradual day to day transition so that there are various intermediate ranges of amplitude or sweep of the tide between one extreme *spring tide* and the following neaptide and *vice versa*. The maximum amplitude of the spring tides is found during the four equinoctial periods of the year.

Between the extreme high tide and the extreme low tide levels numerous levels on the shore are recognized relating to the different phases of the tide. Main levels are: Extreme Low Water Springs, Mean Low Water Springs, Mean and Extreme Low Water Neaps, Mean High Water Springs and Extreme High Water Springs.

The rise and fall of the tide exists in the sea in all parts of the world but the amplitude and the nature of the tide itself may vary to some extent from place to place. In many equatorial coasts the range or the amplitude is very low; but there are other situations elsewhere (e.g. the Bay of Fundy) where the tidal amplitude may reach 50 to 60 feet even. On the coasts of southern India the range (that is, the vertical difference between the highest and lowest tides) is only a few feet, while in some situations elsewhere as along the Gujarat coast the tidal range is quite considerable reaching upto 15 feet and above. In shores with normal slopes a greater amplitude of the tides is associated with a greater area for the occupation of the fauna and flora.

(b) *Substratum*: The substratum of the seashore is made up essentially of mud, sand and rocks. Often, these different kinds of substrata intergrade into

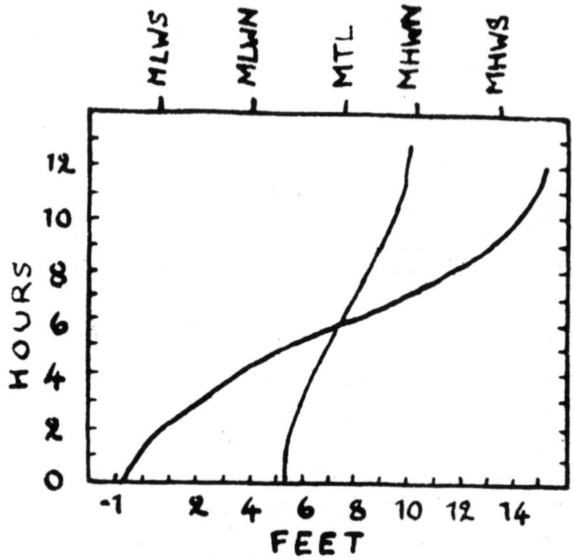

Fig. 5. Duration of Exposure on the Shore at Cullercoats, England during a Spring tide and a Neap tide (Reproduced from Seshappa, 1966)

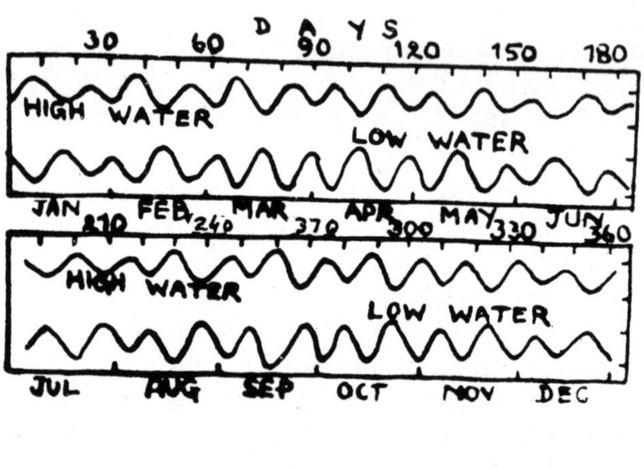

Fig. 6. Diagram of Diurnal Tides for One Year for a Selected Station in England for 1947 (Reproduced from Seshappa, 1966)

one another in different combinations and degrees. In some places there may be only vast expanses of rocky shore while in others there may be vast stretches of sand and in still other places there may be areas of very soft mud or slush which may sometimes be too dangerous to tread on during our work on the shore. The rocks may be broken up into boulders or small stones or shingle; the sand may be coarse or fine; pieces of stone and patches of sand may be found in areas of mud, and muddy islands may be found in areas which are mainly sandy or stony. As the tidal influence is felt on all these different kinds of substrata at all the levels of the shore, it may be said that the environmental factors for the intertidal organisms are changing very frequently, the duration of the exposure to air and light being also affected among other factors of the area. Naturally, the organisms of such an environment need to be adapted to the difficulties of the situation and it may be stated that the struggle for existence among these organisms in the area is very severe indeed. This struggle for existence of the intertidal organisms is on the one hand between the organisms and their environment, and on the other among the organisms themselves (among one another). In spite of the harsh nature of the surroundings however, there is no scarcity of animal or plant life on the seashore.

(c) *Osmoregulation*: A special feature of most animals of the shore and the shallow seas, as in estuaries, is that of a very high ability of osmoregulation, that is, the ability to adjust the concentrations of certain salts in their body fluids to suit their concentrations in the outside medium in which these creatures live. This capacity of osmoregulatory control is particularly highly noticed in animals of the estuarine environment where the freshwaters of rivers meet and mix with the salt waters of the sea. The adjustment of the salt concentrations is brought about by the osmoregulatory flow of substances through the body membranes. This ability is as a rule limited or absent in the animals of the deep sea and open ocean, and of course the environment there does not seem to necessitate the same.

(d) *Zonation*: When the tides of the sea recede that is, at low tide, the organisms, of the shore are exposed directly not only to the air but also to the sunlight and heat. At the higher levels of the shore the exposure to these factors is relatively for a long duration while at the lower levels this may be only for a negligible or nominal duration; the splash effects of the waves may still wet these levels constantly and reduce the effects of insolation that is, the sun's heat (along with the other effects of exposure). Thus the influence of the factor of exposure is different on the different organisms at the different levels of the shore. In adaptation to these differences we find in this environment , different groups of organisms and combinations of various organisms forming different zones of the shore. As a result of sustained researches in different parts of the world's seashores, it has been found that the intertidal organisms of the world in general can be divided into three distinct zones of organisms; one of these is the zone near the low tide level; another is the zone near the high water level that is *exposed to air* for relatively long intervals of time between successive tides and wetted only for short periods; and the third zone is the intermediate zone lying

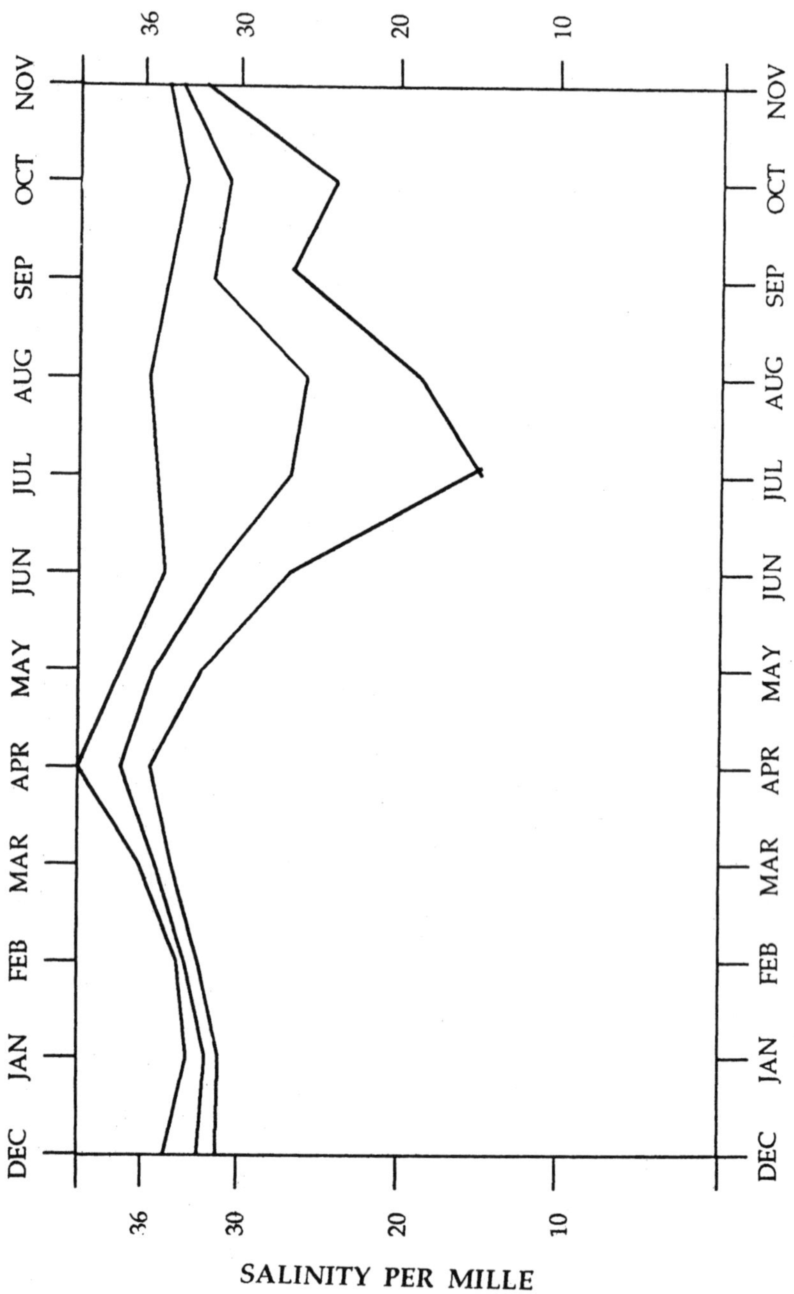

Fig. 7: Monthly average salinities of seawater at West Hill, Calicut during the period Dec. 1949 to Nov. 1950. Middle Curve = Means; Top and Bottom Curves = Extremes (Data from Seshappa, 1953)

between the above two around the mid-tide level forming the life zone of the mid-beach. This statement is however, a very rough generalization only while actually many more zones and subzones may be found on the shore in some places depending on the particular environmental situations encountered from place to place. In each zone the organisms can be noticed to have formed into an *"ecosystem"* adjusted and adapted to all the complications of the environmental factors of the particular area. Another special feature to be mentioned about the shore environment is that most of the animals and plants that are found living here do not seem to be found in any other ecological situation at the present time, if we ignore some of the low-tide level creatures which do extend for a short distance and depth into the sublittoral areas of the sea also as observed at the present time.

(e) *Plant populations*: Near the low tide level of the rocks, it is usual to find some of the larger plants such as *Enteromorpha*,. *Ulva, Chaetomorpha, Padina, Gracilaria,* and *Porphyra*. Very often the plants of the low tide zone form easily recognizable patches on the rock with clear upper level contours which follow in outline the exact line of the waves and tides beating on the rocks at particular states of the tide. Some of the plants of the seashore are useful as food; forms like *Porphyra* and *Gracilaria* are well known for their nutrient values. A food item called *Agar* is manufactured from *Gracilaria;* different species are said to yield agar ranging between 12% and 43% of the plant weight. *Porphyra* is said to contain 30 to 40 per cent protein by weight and thus it seems to be one of the most important plants among our marine species from the commercial point of view. Useful algal species are found in different parts of the Indian coast such as, Mandapam, Pamban, Rameswaram, Kilakarai, Tuticorin, Cape Comorin, Muttom, Colachel, Kovalam, Vizhingam, Varkala, Karwar, Ratnagiri, Bombay and Dwarka on the west coast, and Visakhapatnam, and Chilka lake on the east coast. Most of these plants do not grow in areas of sand and mud in general as a rocky substratum is necessary for most of them for their good growth and survival. There are however, a few forms which can live in non-rocky areas of the shore also.

(f) *Animals: Mortality of Emerita asiatica*: The intertidal zone provides a suitable niche for various kinds of animals which may live sheltered on or under the algae or directly attached to the rocks and stones or even creeping about on the surface layer of the mud and sand or burrowing into it. Many of the burrowing animals carry out the feeding and other activities when submerged in the water; during exposure to the air (i.e. during their emergence outside the water because of low tide conditions) these animals may remain hiding inside their burrows. A very good example of a similar mode of living is seen in the case of the common mole crab, *Emerita asiatica* of the sandy areas of our coasts at low tides in some places. This animal normally cannot live at the higher levels of the shore and is a sand-dweller not able to live well in very fine slushy mud also. There was an instance at Calicut several years ago as observed by the present writer (Seshappa 1953) when owing to some peculiar happenings in the environment the sea-water near the shore became a thick slush of mud rising up on the shore along with the rising high tide, instead of the clear water of the sea; the low water region was normally of fine sand with a dense population of the above mole crabs. When the mud rose up on the shore along with the high tide, the mole crabs faced suffocation; being unsuccessful in their attempts to move up to the higher levels of the shore for escaping from the slush and its suffocating effect, they finally died in large numbers; a vast expanse of the dead creatures was found around the mid-tide level when the tide receded in its turn. The mole crabs were also physically unable to swim out of the slush, their organs of locomotion being perhaps not strong and equal enough for the situation. The destruction of these creatures on the shore was so vast and extensive at that time that it took several years altogether before any considerable recolonization of the area by the species could be noticed again. The mole crab is not an organism of any known economic importance at present but other organisms which are of direct importance to man may also be affected by such happenings. Over a period of years, however, in the normal course of events, the natural forces of the biological cycles in the sea assert themselves and usually help in the re-establishment of the original conditions of what may be called the balanced populations of various organisms that establish themselves in any given area. The movements of the eggs and larval forms, the drifting of organisms with currents and wind waves and other such homogenizing forces all help in the eventual return of the ecological balance. This does not again mean that the same populations or groups of species may re-appear in the same numbers and in the same locality always after the recovery following such destruction or mortality. On the other hand all such communities may show a trend towards a change in one direction or the other, the environmental conditions acting at the time adding the deciding weight to influence the organisms one way or the other. In the intertidal region, even when the range of the tide is very small, say about one foot only, the conditions of life even within the narrow strip of the shore can be very different at its different levels, these showing a sort of telescoping effect when compared to regions of great tidal amplitude. The ability to adjust themselves to the variations in tidal influences varies among organisms from species to species and race to race. Because of this, different levels of the shore have different genera and species of plants and animals. It has also been noticed often that not only different species of the same genus but also different subspecies, races or varieties of the same species and different genera of the same family can occur at different levels of the shore forming clear zones distinct from the others. Some mobile animals have the ability of returning to their original zone of existence even when removed and placed away from their original niche or site.

(g) *Speciation*: Algae of the genus *Fucus* grow abundantly on the intertidal rocks in the British Isles and the North Sea. In this genus, *F. serratus* occurs at the lowest level, *F. spiralis* at the highest level and *F. vesiculosus* in the middle levels of the shore, *Pelvetia canaliculata* also occurring along with *F. spiralis* at the highest levels. As Zaneveld (1934) has shown, these three species of *Fucus* have quite different degrees of transpirational activity or the ability to withstand desiccation by retaining moisture within themselves for relatively long periods, nearly or exactly in relation to the levels of their occurrence on the shore. Among the animals the prosobranch mollusc *Littorina* is an example of a genus with different species living at different levels of the shore, the genus itself being represented all over the shore from the lowest to the highest regions. Of these, *L. neritoides* lives at the highest level including the spray zone, while *L. littorea* occupies the lower tidal levels. A third species *L. littoralis* (=*L. obtusata*) occurs in a zone intermediate between the above two and is abundant around the midtide level in the shelter of the *Fucus* plants and also on open rock. These species of *Littorina* have three different kinds of graded reproductive methods in that, *L. littorea* releases planktonic larvae, *L. littoralis* lays eggmasses on rocks and weeds, and *L. neritoides* is viviparous producing shelled juveniles from the "uterus". Another species of *Littorina* called *L. saxatilis* has three or four subspecies which occur at different levels of the shore just as the different species of the genus *Littorina* themselves occur at the different levels, a zonational pattern being shown by both.

(h) *Sessile and non-sessile forms*: Among the animals living attached to the rocks in the intertidal region are limpets, barnacles, sponges, hydroids, polyzoans, and several other groups of animals of the Invertebrata. The Actinians or sea anemones are simple tubular animals of varying sizes attached to rock at one end and having the mouth surrounded by whorls of tentacles at the free end. Some of the anemones are quite beautiful with pleasing and attractive colours and patterns particularly on the tentacles. During the low tide when the creatures emerge from the water the resultant exposure to sun and air makes these animals contract a great deal so that their form cannot be fully recognized until they expand again. They expand their bodies and tentacles when they are again covered by the water of the rising tide, and they then appear blooming out as before. There are nematocysts or stinging thread cells on the tentacles and these help in offence and defence bv stinging the creatures that may come into contact with them.

The limpets (Fig. 27) are gastropods of the group Prosobranchiata and nave very limited powers of locomotion, the animals remaining encased in a flat-cone-shaped shell which remains firmly attached to the rock.

The barnacles (fig. 30) (e.g genus *Balanus*) belong to the class Crustacea of the phylum Arthropoda or the joint-footed animals. They live attached firmly and permanently to rock; their soft parts are contained and protected in a shell of calcareous matter made up of several hard, thick and ash-white opaque pieces moveably held together so as to fully enclose the soft parts of the animals in the closed condition. When submerged in the seawater, the shells open out and the appendages can be easily seen waving about outside the shell and capturing the food organisms. Under exposure to sun and air that is, when the low tide leaves them high and dry on the exposed rock, the shell valves close up and protect the animal against desiccation and attack by predators.

Amongst the non-sessile animals of the shore but with only a limited locomotory ability, may be mentioned various isopods, amphipods, polychaetes, gastropod molluscs, starfishes (fig. 28), brittle stars (fig 29), hermit crabs and crabs (fig. 55) etc., some of these being relatively more agile than the others. There are usually also various sizes of pools and puddles on the intertidal rocks and these too provide shelter; various creatures take such protection by getting into these pools and puddles during low tide which other-wise exposes them to the air and sunlight which they cannot tolerate beyond certain limits. Some of these pools have also their own locally developed communities of plants and animals all grown in association with one another and living together though not necessarily symbiotically in a strict sense. Again the pools at different levels of the shore may have different communities altogether, in keeping with the nature of the surrounding fauna and flora of the intertidal rocks themselves.

The hermit crab is a unique type of animal living inside the empty shell of certain gastropods. Because of this habit the body of this animal is much modified, some of the appendages being reduced or lost. The shell has to be carried about by the hermit crab wherever it moves about; though this is certainly an encumbrance, the shell provides the most efficient protection for the otherwise endangered delicate creatures when they are confronted with predators. On such shells it is also sometimes seen that a sea anemone (which may be said to be even much more delicate than the hermit crab) settles down and grows up living permanently on the shell surface. This forms a good mutually beneficial association in view of the efficient stinging cells of the sea anemones being also available for the group for co-operative defence. The movements of the shell as carried out by the hermit crab inside it, are sufficient for the sea anemone for its own life activities.

(i) *Sex Change in Patella vulgata*: Some animals of the intertidal region show some very interesting peculiarities in their biology. There is for instance the limpet called *Patella vulgata* of the rocky shores of Great Britain which has such features; this animal is a gastropod with a flattened cone-like shell as stated above, which fully covers and protects the animal within itself and remains applied firmly to the rock by its wide-open mouth. The limpet is abundant around the mid-tide level (or mid-beach as it may be also called) but its occurrence extends to some height above as well as below the mid-tide level, stray individuals however occurring at the highest high tide level and even beyond, and also down to the lowest low tide level. When the size-distribution of this animal was studied at Cullercoats (Das and Seshappa 1947) it was found that the smaller sizes were more numerous at the lower levels of the shore while the larger sizes were more numerous at the higher levels of the shore; then among the smaller sizes and at the lower levels of the shore, the males were more numerous while among the larger sizes and at the higher levels of the shore the females were more numerous. This indicated the probability that the gonads may be changing in many specimens with increase in size and age which is also correlated with the upward migration on the shore. There are certain species of oysters which show such a sex change quite regularly.

(j) *Sand and mud forms*: In coarse sand and gravel the type of animals found on the solid rocky shore are rare. In fine sand some water can be retained even during low water and some animals remain buried inside the sand and thus sheltered and wetted even during the low tide periods when they are exposed to sun and air. The mole crab *Emerita asiatica* mentioned earlier is an animal normally restricted in its occurrence to the fine sand at and near the low water level. A bivalve mollusc called the wedge clam (*Donax cuneatus*) also lives in the same situation (practically sharing the same *niche*) with the mole crab mentioned above, but sometimes all by itself. Several species of animals inhabit the sandy zone also as well as the gravel and shingle and these situations are also quite favourable to these creatures. Some crabs burrow holes on the sand or occupy holes already made and left by other animals; some of the echinoids are also known to show this habit among other animals.

Many species of worms especially the polychaetes, occur in sand and in the sandy and muddy patches below stones. A species of Polychaeta called the lug worm (*Arenicola marina*) was not known to occur around our country some four decades ago though it was abundant in most other areas including also the North Sea. In the early fifties of this century the occurrence of this worm was discovered by Ranade (1952) in the sands off Bombay and worked out by Kewalramani et al (1960); more recently the species has been reported from other areas of our coasts also. As our seas and coasts were much less explored than the coasts of many of the more advanced countries, it is not at all surprising that many species remained and are still remaining undiscovered and unworked out. Actually new species are coming to light in most groups of the animal as well as the plant kingdoms with the further progress of exploration and research work in the different areas. The polychaete fauna of our shores had been worked out to a considerable extent even by 1932 by P. Fauvel (1953) though in this group also many species are still coming to light at the present times. Olga Hartman (1974 a & b) has given an uptodate position regarding the study on polychaetes of the Indian Ocean and other regions with extensive bibliography.

The polychaetes are worms of the group Annulata to which the earthworms and leaches also belong. The name polychaete signifies the presence of bundles of stiff hair-like setae, these projecting out of the parapodia at the sides of the segments of the long segmented body. These parapodia are muscular projections on each side of each segment and bearing bundles of the setae in muscular pouches; the parapodia are organs of locomotion and the setae help in offence and defence. The larval stages of the polychaetes (fig. 18 & 19) are planktonic and are provided with locomotory cilia on their bodies. As metamorphosis progresses and the larvae descend to the bottom layers (as many do) to finally settle down as benthic creeping animals, they assume the normal adult structure. However, some polychaetes may spend all their lives in the surface waters only as part of the surface plankton. Some polychaetes secrete protective tubes of mucus, "calcium", chitin, mud, sand etc. on other substrata and live inside them. Fig. 18 & 19 show the larval stages and fig. 56 an adult of polychaetes.

(k) *World trend in Zonation*: The animals of the intertidal region are naturally different on the coasts of different countries in different regions of the same coasts also. But it has been noticed clearly that when the animals and plants of the different zones of the shore are considered in terms of tidal levels, they invariably belong to the corresponding groups of the animal or plant Kingdoms and Orders; the reason for this remarkable phenomenon is of course the similarity of the corresponding environmental factors themselves to a great extent.

(1) *Sea Turtles*: There are some animals that habitually live in the sea but visit the shore sometime or the other during their lives. The marine turtles are an excellent example of such creatures that cannot complete their life-cycles without visiting the shore. Two common examples of Indian marine turtles are: *Chelonia mydas* the green turtle, and *Dermochelys coriacea* the leather-back turtle; other seaturtles of our region are: (1) *Eretmochelys imbricata* (the hawksbill turtle), (2) *Lepidochelys olivacea* (the Olive Ridley) and (3) *Caretta caretta* (the loggerhead turtle).

Sea turtles come to the shore for egg laying only, this landward migration from the sea being for the purpose of breeding or laying the eggs; they migrate from almost deep waters far away from land taking several days sometimes for this movement, and reach the sandy shore having their uteri full with eggs in advanced stages of maturity and ready to be laid; immediately on reaching the sandy shores they proceed carefully to select a suitable site fairly high up in the intertidal region when the tide is down and the sandy stretch exposed. Then with its limbs each animal digs a deep and wide pit on the sand and soon deposits eggs in it in several successive batches. The eggs are much larger than hen's eggs and from a hundred to a hundred and fifty or more of these may be deposited in each laying trip by each turtle, all being done in night, normally. Then the eggs are fully covered with sand; the mother turtle with its breeding mission thus completed, leaves the eggs to their fate and returns to the sea as quickly as possible before daybreak; after sunrise there is always the danger of more enemies in the form of predators and man himself. When the eggs laid and covered inside the pit hatch out, helpless little juvenile turtles wriggle out to the open shore and instinctively move towards the seawater to start their sealife. Many juveniles are almost always caught out by enemies and eaten up or destroyed on the beach before they can reach the sea which is their home.

Turtle eggs are a common and popular item of human diet in many countries of the world; as such they fetch a good price. The large numbers of eggs that are laid in each spawning makes the search for these eggs a paying pastime as well as a profession for the fisherfolk concerned. Turtle culture is also in vogue in some places along our coasts both for the flesh and the eggs of the creatures. Turtle soup is a delicacy in some countries.

In India there is a good turtle fishery in some places and it is especially abundant along the Orissa coast and also on the Tamil Nadu and the Kerala coasts.

The Karnataka Government also seems to be planning a turtle culture farm in a big way somewhere along its shore on west coast, according to newspaper reports.

Chapter VI

The Plankton

(a) *Introduction:* The floating and drifting organisms of the sea which have little or no independent movements of their own form the plankton. These include both animals and plants; the most minute organisms which can be collected only by centrifuging the seawater and that cannot be collected by nets in the usual way (even the finest meshed net being incapable of capturing all of them) are called the *nannoplankton.* The nets used in the collection of ordinary plankton are described in another chapter later (see p. 132); as stated already the plants of the plankton are together called phytoplankton and the animals are together similarly called the zooplankton. Most of the animal groups of the world are represented in the sea; in some stage or the other most of them find a place in the plankton also. For mankind the fishes may be said to be the most important marine animals from an economic viewpoint; these (fishes) depend on the planktonic organisms for their livelihood, both zooplankton and phytoplankton being used as food by them as a group. Planktonic organisms of particular kind or composition, may attract particular kinds of fishes to a given locality in the sea; it is known also that when a particular organism or group of organisms becomes scarce in a given mass of water or these are not produced in the required quality and quantity the fishes which feed on them normally also fail to turn up in the usual way, this resulting in the failure of the fishery. This is one of the important explanations for the frequent fluctuations found in the herring fisheries in the North sea and probably similar is the case to a large extent with the mackerel and oil sardine fisheries of our west coast.

(b) *Copepods and other forms*: Among the various groups of planktonic animals, the copepods are about the most important in the food of fishes in general. The copepods occur as adults as well as larvae and both the stages are profusely eaten by various fishes. Other groups of importance as food of fishes and occurring in the plankton are the cladocerans (e.g. *Evadne* and *Podon*), various Decapod larvae, various molluscs in adult as well as larval stages, tunicates, *Noctiluca,* fish eggs and larvae, polychaetes, arrow worms etc. Mukundan (1967) studied the inshore plankton at Calicut during the years 1957 to 1965 continuously and this work can be cited as an example of general studies conducted on plankton both for fishery biological assessments and academic speculation. His findings provide a reliable picture of the plankton conditions in those years on the west coast in general (with reference to Peninsular India) in

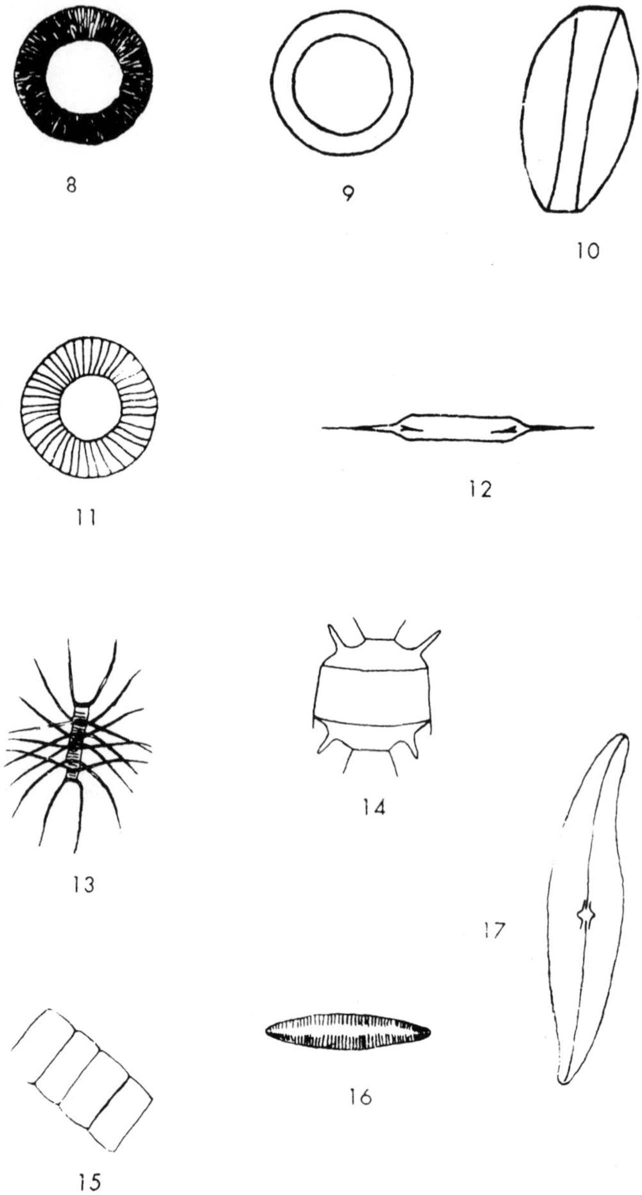

Fig. 8. *Cyclotella*, Fig. 9 & 10. *Coscinodiscus* (Two Views), Fig. 11. *Planktoniella*, Fig. 12. *Rhizosolenia*, Fig. 13. *Chaetoceros* (A Piece from a Chain of Cells), Fig. 14. *Biddulphia* (One Cell), Fig. 15. *Fragilaria* (Piece of a Chain of Cells) Fig. 16. *Fragilaria* (One Cell), Fig. 17. *Pleurosigma*

the light of the earlier findings of George (1953) who worked in the same area and was able to prepare a calendar for the plankton on the basis of a thorough study over some earlier years.

Mukundan found that the copepods formed the single consistently dominant group in the zooplankton, they also being the most abundant in the year taking the zooplankton as a whole; they were absent during the southwest monsoon and showed peak conditions of their occurrence in October-November period, this status being retained with slight variations till the following monsoon. Cladocerans occurred in almost all months with their peak occurrence during the July-October period. Decapods and their larvae, Chaetognaths, polychaete larvae, molluscan larvae and *Noctiluca* were the other important forms in the zooplankton. In almost all the years there was a rich crop of polychaetes and their larvae in the inshore plankton, this starting when the peak of the monsoon was over and lasting for three to four months thereafter. The majority of these were spionid larvae but polynoid, chaetopterid, Nereid and Magelonid larvae were also included. Both gastropods and bivalves were well represented of the molluscan larvae, the bivalve larvae being however numerically more and generally occurring from November to February. The gastropod larvae had a similar pattern of occurrence but they were generally abundant in the summer months. Of the pelagic gastropods both pteropods and heteropods were represented, but the pteropods were generally more abundant; *Creseis* was the main genus with *C. acicula* as the most common species. *Atlanta* was the main Heteropod genus. The tunicates were represented by the Thaliacians like *Pegea confederata* and *Thalia democratica,* and also by the Appendicularians like *Oikopleura and Fritillaria.*

Fish eggs and larvae also form a significant part of the plankton, and especially so during the August-December period in the inshore plankton. Mukundan found Carangid and Clupeid eggs dominant among these, significant numbers of *Cynoglossus* eggs also being however present.

(c) *Fluctuations*: The pattern of fluctuations in the west coast plankton as seen at Calicut can be divided into two categories: (i) including most of the groups and declining sharply after the onset of the monsoon rains which lower the temperature and salinity to a marked extent, (ii) mainly the cladocerans *Evadne* and *Podon*, and also *Noctiluca* mentioned earlier elsewhere. These swarm during monsoon months also but often with abrupt appearance and disappearance. George (1953) prepared a monthly calendar for use round the year for both zooplankton and phytoplankton at Calicut but mostly applicable to the southern region of the west coast consisting of the Kerala and Karnataka coasts; he found for three years (1948-51) that the total plankton volumes were very high during the southwest monsoon months and especially in the early half of that period, and that the dominant factor responsible for these high values was the swarming of these diatoms and dinoflagellates; the postmonsoon season during September to December was characterized by the re-appearance in large numbers of copepods, cladocerans, planktonic decapods and fish eggs and larvae with moderate numbers of diatoms and dinophyceae.

Mostly the same or similar forms of both zooplankton and phytoplankton occur on the east coast also (Chacko 1950, Prasad 1952, Prasad and Nair 1960, Prasad 1956 etc), the main environmental difference closely associated with the plankton cycle in the sea between two coasts being the absence of the southwest monsoon on the east coast, this playing a very dominant role in the biological cycle in the sea on the west coast; on the east coast on the other hand, there is the northeast monsoon which occurs over a different period of the year (i.e. November to January) and is slightly of a shorter duration. The majority of the copepod species seem to be the same in the Palk bay and the Gulf of Mannar (Krishna Kartha 1959); maximum breeding of the copepods was noticed during the May-September period in the Palk bay but during the September-March period in the Gulf of Mannar. Decapods, Lamellibranchs and gastropods formed the most important groups (altogether forming about 60% of the total) of all the larvae, the major elements varying from year to year to some extent (Prasad 1954); according to the plankton calendar of Prasad (1958) for the Mandapam area, the total plankton production was maximum in March in the Gulf of Mannar but it was only in June in the Palk bay. The annual cycle showed two production peaks in the Gulf of Mannar but only one peak in the Palk bay; phytoplankton was highest in May-June in the Palk bay but only in October in the second half of the year in the Gulf of Mannar.

(d) *Phytoplankton work*: One of the most outstanding contributions to phytoplankton studies in recent decades in the Indian inshore sea areas may be said to be that of Subrahmanyan and his colleagues (1946, 1959 1960 and 1969). Subrahmanyan described over 500 species of the phytoplanktonic forms of all groups together representing over 150 genera from both coasts of India; in 1946 he found 171 forms of marine planktonic diatoms at Madras, these representing 15 families, 64 genera, 134 species, 17 varieties and seven other forms; about 50% of the total forms were common to Indian seas and the Java sea. Nair and Subrahmanyan (1957) found a close correlation between the abundance of *Fragilaria oceanica* and the good fisheries of the oil sardine *Sardinella longiceps* on the west coast. Subrahmanyan and Sharma (1960) listed a total of 291 species of phytoplankton from Calicut alone these including among others, 29 species of Bacillariophyceae, seven species of Dinophyceae, and one species of Cyanophyceae which are all abundant and of special importance. The "swarming" forms of diatoms (the Bacillariophyceae) include *Asterionella japonica, Bacteriastrum hyalinum, Biddulphia heteroceros* and many species of *Chaetoceros, Coscinodiscus, Fragilaria oceanica, Nitzchia sigma, Rhizosolenia* spp *Skeletonema, Thalassiothrix,* and some other forms. Among the Dinophyceae the "swarmers" include *Dinophysis caudata, Ceratium* spp, *Glenodinium, Peridinium* and *Noctiluca miliaris*.

The diatoms form the bulk of the marine phytoplankton. Their maximum production on the west coast takes place during the southwest monsoon period (June to September); a second peak in the production but of a minor importance occurs later between December and February, these two blooms of plankton being comparable to the spring and autumn blooms of the temperate seas; this

Some Zooplankton Elements (I)

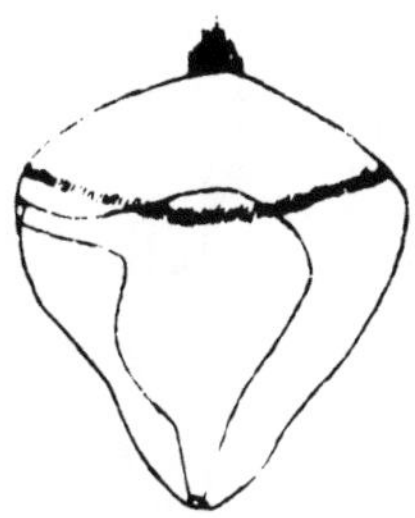

18

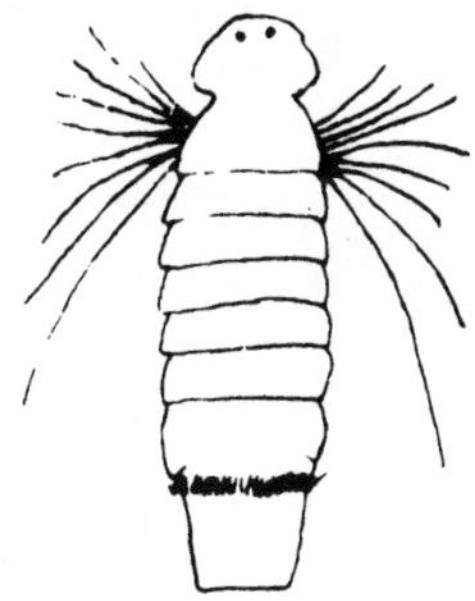

19

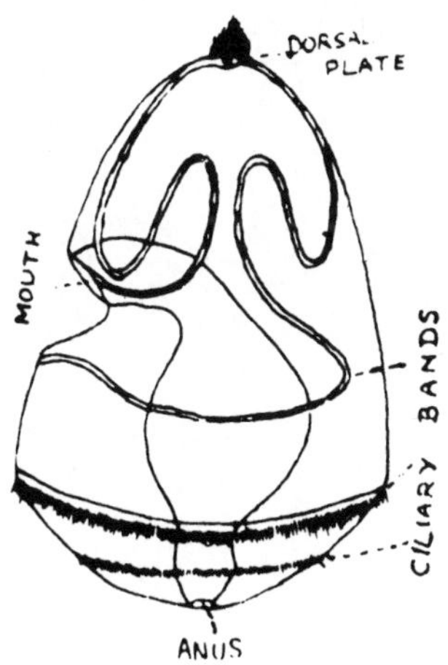

20

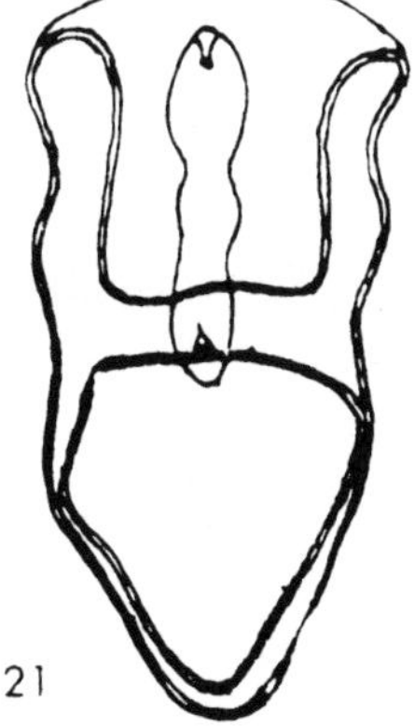

21

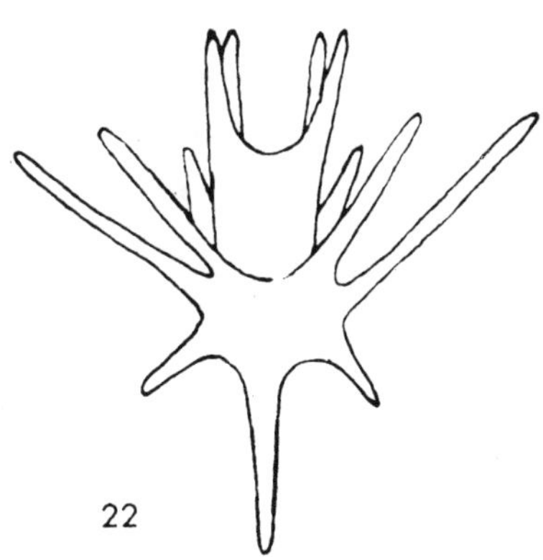

22

Fig. 18. *Trochophore* Larva of a Polychaete, Fig. 19. Advanced Larva of a Polychaete Fig. 20. An Advanced *Tornaria* Larva of *Balanoglossus*, Fig. 21. Early *Tornaria* Larva of *Balanoglossus* Fig. 22. *Echinopluteus* Larva of a Sea urchin

Some Zooplankton Elements (II)

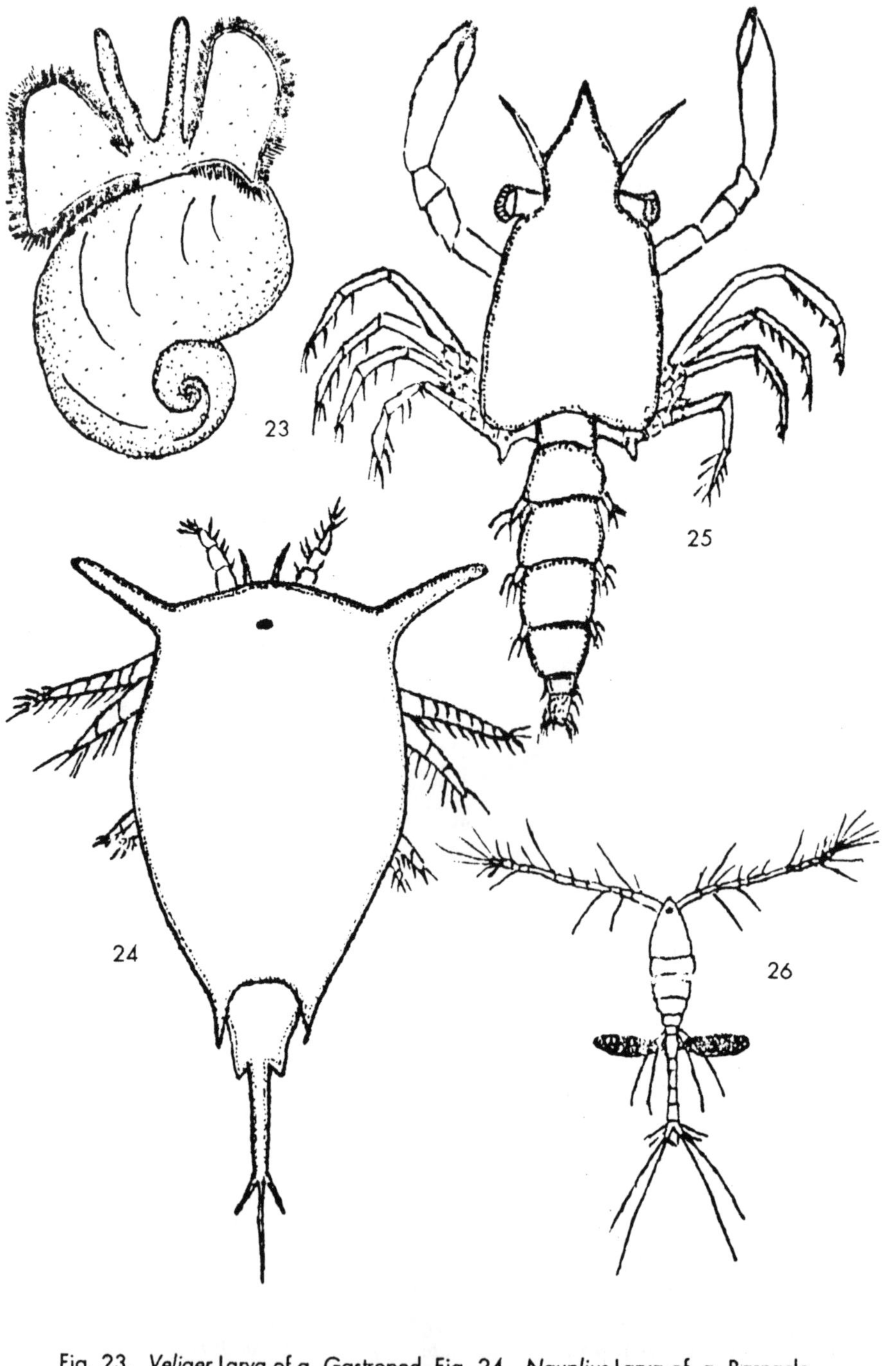

Fig. 23. *Veliger* Larva of a Gastropod, Fig. 24. *Nauplius* Larva of a Barnacle
Fig. 25. *Megalopa* Larva of a Crab, Fig. 26. A Copepod (Adult)

bloom of the phytoplankton of the southwest monsoon period along the west coast surpasses in quantity similar blooms known in some of the most fertile areas of the world.

On the east coast also the plankton is quite rich in variety as well as bulk, most of the species being of a cosmopolitan distribution but as stated above the periods are different from the west coast. The zooplankton and phytoplankton relationships are generally inverse in the Gulf of Mannar but often direct in the Palk bay. The diatom communities of Mandapam and neighbourhood seem to fall into the following biogeographical categories according to Prasad and Nair (1960): Arctic, North Temperate Neritic, Temperate Oceanic, Tropical Oceanic and Littoral (Unclassified); these authors have listed 27 neritic, 13 oceanic and nine littoral forms and also two unclassified species. "Out of the several species only 15 were well represented, most of these exhibiting some kind of periodicity. The principal forms were pelagic neritic and oceanic species."

(e) *Role of Phosphates*: The production of phytoplankton is closely related to the nutrient cycles in the inshore waters as shown by George (1953) on the west coast. He found a marked outburst of diatoms and dinoflagellates, high total plankton volumes, appreciable lowering of water temperature and salinity, and high phosphates to be characteristic of the southwest monsoon period. The dominant diatoms noticed by him were *Fragilaria oceanica, Coscinodiscus gigas,* and species of *Rhizosolenia* and *Chaetoceros,* while *Bacteriastrum, Skeletonema, Eucampia* and *Asterionella* were also present. Continued studies for three years showed that the production of diatoms increased and decreased frequently (and in any case more than once or twice during each year) there being also subsidiary fluctuations within each production peak period. These ups and downs in production were explained by Jayaraman and Seshappa (1956) as being correlated with the ups and downs of the phosphate values in the surface water of the inshore sea area. Phytoplankton blooms were closely related to the increase of the nutrient salts like phosphates, this occurring immediately on the commencement of the southwest monsoon on the west cost. Jayaraman and Seshappa found that there were waves of production and decline in the nutrients as well as in the phytoplankton elements in the same season but each following the other as follows:- (1) the first diatom bloom occurs after the increase in the values of the nutrients, especially of the phosphates and nitrates; (2) the diatom values decline subsequently with a falling trend of phosphates which are utilized by the bloom of the diatoms; (3) but the phosphate and nitrate values increase again with their regeneration from dead phytoplankton as well as replenishment from the seabottom; (4) this sequence continues; such increase and decrease of the phosphates first and then of the phytoplankton occurs more than just once or twice during the same season; the phytoplankton is grazed by the zooplankton and other animals which become dominant in the sea immediately following the cessation of the southwest monsoon.

Chapter VII

Productivity

(a) General

The term productivity is used for the total biomass produced by the different bio-ecological groups of organisms in the sea. The total biomass of phytoplankton provides a measure of the primary productivity; the zooplankton provides a measure of the secondary productivity, while the fishes and other large animals which grow by consuming the primary and secondary items produced in the sea form the tertiary productivity; the total fish and other useful animals of the sea are all products of tertiary productivity. In the earlier years of work on the productivity, figures for the sea were available mostly only for the phytoplankton and also only in terms of dry and wet volumes of the plankton and of the total numbers of cells in the same as estimated from samples. Pigment extracts were also taken separately, of the net phytoplankton and nannoplankton, these together giving the pigment values. The nannoplankton was collected by re-filtering the samples in a 60u net after passing it through a number 16 bolting silk net. A study of over eight years (1955-72) by Subrahmanyan and Vishwanatha Sarma (1965) from three stations near Calicut (4 fathoms, 10 fathoms and 20 fathoms) showed a peak production during the south west monsoon every year. The highest monthly average in this work was 1,71,000 pigment units (PPU/M^3) noticed in September 1956, and the lowest was 1000 PPU/M^3 noticed in January 1958; the absolute range from individual readings was 400 PPU/M^3 to 24800 PPU/M^3. The nannoplankton was found to be forming 85% of the crop in August 1960, 84% in June 1959, and 96% on 8-7-1969; the nannoplankton was thus found to be quite important and it generally formed between 30% and 50% of the standing crop with an average of about 40%.

Considerable work has been done on primary and secondary productivity in Indian waters by Nair (1970), Pillai and Nair (1972), Yonge (1972), Tranter and George (1972), Prasad and Nair (1973), Qasim (1977 & 1979), Qasim and others (1978), and Ramachandran Nair and Pillai (1982), in addition to Subrahmanyan's pioneering work earlier. Qasim has reviewed the biological productivity of the Indian Ocean including the Arabian Sea as well as the Bay of Bengal.

(b) Primary production

The total surface production of upto one metre depth in the Bay of Bengal (within the area of 40.87 x 10^5 KM^2 (lat. 0° - 25°N and long. 80° - 100°E) has been worked out to be 20.4 x 10^6 tonnes of carbon (C) per year giving an average of 4.9 tonnes of carbon per KM^2 per year. The areas of maximum production are in the coastal waters i.e. coast of India, areas adjacent to Sri Lanka, Visakhapatnam and Burma. On the west coast also the total surface production is high, the average for the Arabian Sea being lower than that of the Bay of Bengal by one tonne C/ KM^2 per year. The reason for the high production in the Bay of Bengal may be mainly the run off of nitrogen and phosphorus from rivers—an annual average of 1300 KM^2 giving rise to this run off. Adding the run off from Bangla Desh and Burma this run off would be about 2000 KM^3 . A more important reason given is the greater cloud over the sea in the Bay of Bengal than in the Arabian Sea where the sky is cloudy only during the southwest monsoon.

The productivity in the column water is the reverse of the trend seen at the surface. The Arabian Sea is far more productive than the Bay of Bengal, the regions of maximum production being the Gulf of Oman, around the mouth of the Indus river, the Saudi Arabian waters, the waters of the Mosambique coasts and the west coast of India; the high production noticed in these waters is mainly due to upwelling. The total production of the column water in the Arabian Sea has been worked out to be 1064 x 10^6 tonnes of C per year; for the entire area of the Indian Ocean covered by the Indian Ocean Expedition the total column productivity has been estimated as 4.42 x 10^9 tonnes of carbon per year and 46.9 x 10^6 tonnes of C per KM^2 of the sea. For the Bay of Bengal area alone the productivity is 46.9 x 10^6 tonnes C/year. For the entire Indian Ocean the average primary production rate has been estimated as 94 tonnes of carbon per KM^2 per year or 258 mg C/m^2/day (Qasim 1962, p. 129). Dr. Qasim after a comparative study of the productivity of the world oceans, states that the productivity of the Indian Ocean is almost 75% higher than that of the Pacific and approximately 17% greater than that of the Atlantic.

(c) Secondary production

The areas close to Saudi Arabia, Somalia, and the Northern Arabian Sea have shown the highest biomass values, the other regions of high biomass being the southeast and the southwest coasts of India, and the area adjacent to Sri Lanka upto the Equator. The Bay of Bengal and the southwest coast of India have a uniform rate of secondary production of 5-10 mg C/m^2/day. The pooled values of secondary production for the entire Indian Ocean amounts to 69.27 x 10^6 tonnes of carbon per year (= 1.5 t. CKM^2/year which is equal to 4mg C/m^2/day).

(d) Tertiary production

This is the standing crop or stock from which the "potential yield" or "annual sustainable yield" can be determined. After a detailed discussion of the latest data available at the time on the primary, secondary and tertiary productivity of the Indian Ocean, Qasim (1977) concludes as follows: "Recent figures of pelagic, demersal and crustacean resources from the different regions indicate that the annual catch from the Indian Ocean is about three million tonnes. Hence to bring the catch close to the potential yield, nearly 3-4 times increase is possible annually."

Among other recent works on productivity in the sea as relevant to our waters may be mentioned those of Yonge (1972), Nair and Pillai (1972), and Tranter and George (1972). Yonge studied some aspects of productivity in coral reefs and stated: "Basic productivity of a coral reef must be considered from two aspects: first there is maintenance and growth of the carnivorous Scleractinia (hermatypes and the ahermatypes) which feed on zooplankton. This represents tertiary productivity. Second is rapid growth and maintenance of the reef framework made possible by the imprisoned phytoplankton. This is directly related to secondary productivity". Nair and Pillai (1972) included three reefs in their study, namely, the Manauli reef in the Gulf of Mannar, the Minicoy reefs in the Lakshadweep area and the Andaman reefs in the Bay of Bengal. Primary productivity was estimated by diurnal changes of oxygen in the seawater flowing over the reefs, by using the C-14 technique; they found that the Andamans reefs were non-autotrophic while the other two were autotrophic; "The production of carbon amounts only to 3.9 g/C/day which does not meet the respiratory requirements of the reef organisms" in the Andamans area. Manauli reef produces 2500g/C/m^2 /year and Minicoy reef produces 3000 g/C/m^2 /year (net). These authors state: "Though the Andamans reefs produce 1200 g C/m^2/year the respiratory requirements of the reef organisms is much higher. Hence it is not self-supporting. But the level of community metabolism in the adjacent Nicobar reefs would be more akin to that of the Minicoy reef and so these reefs would be contributing to the economy of the Andamans seas". Bhattathiri and Devassy have found the average productivity in the Andamans sea to be 5.3, 12.4 and 2.4mg/C/m^3/day in the premonsoon period, the column values varying from 0.12-0.62 to 0.20-1.2g. /C/m^2/day in 1979 and 1980. According to Qasim (1962) the tertiary productivity estimated from primary and secondary production rates for the entire Indian Ocean is of the order of 17 million tonnes of carbon per year, this being the approximate potential exploitable rate. He suggests that the conversion of primary to secondary and the latter to tertiary is only in the order of much less than 10% (transfer coefficient). According to Ramachandran Nair and Pillai (1982)" it has now been fully established that the annual sustainable yield for 70% of the Indian Ocean area should be at least 11-12 million tonnes at the present level of world fishing. The EEZ (i.e. the "Exclusive Economic Zone") comprising just over two million Km^2 could provide an annual sustainable yield of 4 to 5 million tonnes." Dr. Swaminathan (1983) had assessed that the total potential fish catch from our own seas is 2,500,000 tonnes out of the Indian ocean total of 7,246,000 tonnes annually, our present catch being only 1,150,000 tonnes, which it should be possible to increase, particularly from tunas.

Chapter VIII

A Glimpse of the Main Animal Groups in the Sea

Introduction

Leaving out a few notable exceptions, practically all the major groups of the animal kingdom are represented in the sea by some example or the other; some groups are however more abundant than others while there are also a few others that have their representatives only in the sea and nowhere else. The more important features of the selected main groups or *phyla* of the marine animals will be briefly mentioned in the following pages:

(a) The Protozoa

The single-celled animals of this phylum are abundantly represented in the sea just as in the freshwaters. Protozoans are found at all depths of the water in the sea, on the seabottom below all water and also on all types of shore such as rocks, mud and sand; they also live attached or settled on other living and dead animal and plant forms. Four important classes of the Protozoa are recognized, namely the Mastigophora or Flagellata, the Sarcodina or Rhizopoda, the Sporozoa and the Ciliophora. The dinoflagellates which form important organisms in the plankton are classified under the Mastigophora while Foraminifers and Radiolarians belong to the Sarcodina (or the Rhizopoda). The Foraminiferans are characterised by the protoplasm being enclosed in a minute shell provided with a large number of tiny apertures through which the protoplasm of the animal flows out from inside the shell forming a network of anastomosing threads on the surface of the shell. Fossil Foraminifers of many species are abundant in many parts of the seabottom, the group being quite an ancient one. A large number of living Foraminifera have also been described from the Indian seas (Gnanamuthu 1943 etc., and Kameswara Rao 1972 etc.). In the living condition the Foraminifers can be collected from all different levels of the seawater, many species being pelagic in habit. In all cases the shells drop down to the seabottom after the death of the animal, the sediment formed by concentrations of such dead shells in some places being called the "Foraminiferan ooze".

27

28

29

30

31

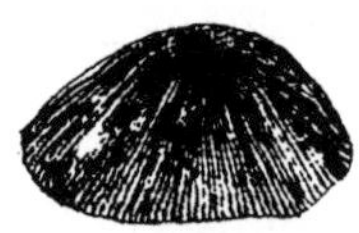

32

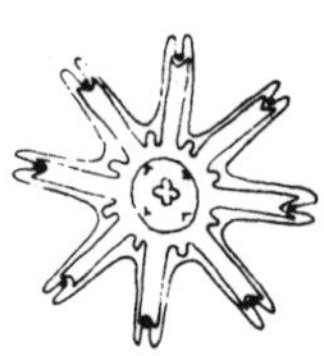

33

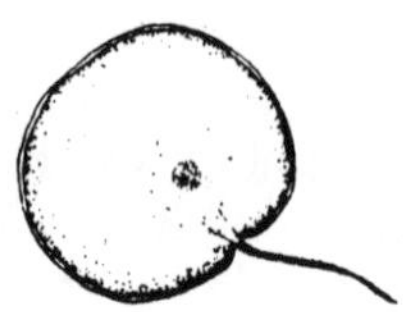

34

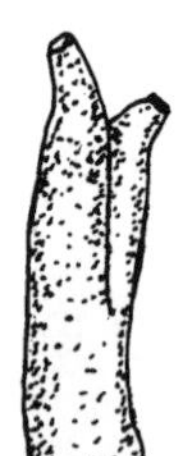

35

Fig. 27. A Limpet, Fig. 28. A Starfish, Fig. 29. A Brittle Star, Fig. 30. A Barnacle (Shell Closed), Fig. 31. *Dendrophyllia* (A Coral), Fig. 32. *Fungia* (A Coral), Fig. 33. *Ephyra* Stage of a Jellyfish Fig. 34. *Noctiluca*, Fig. 35. An Ascidian

The Radiolarians also consist of single-celled structure, each animal being provided with skeletal shell. Like the Foraminiferan ooze is also the Radiolarian ooze occurring in some situations at the seabottom and is constituted by the dead skeletal shells of radiolarian animals.

The Flagellata are free-swimming animals with their cells being not provided with any skeletons, but these creatures retain their body shapes in a manner characteristic of each species by a hardening of the outer layer of the protoplasm into a clear body-wall membrane. These animals have certain whip-like processes projecting out from one end (the "oral" end) of the body and helping in their movements in the medium of water. They also possess the green pigment chlorophyll inside their protoplasm. Because this presence of chlorophyll is very characteristic of the plant kingdom, the Flagellates are partly considered as plants also by the Botanists. But there are many animal characters in the Flagellate structure and hence the Zoologists have not left them out either.

There are numerous species of Flagellates in our coastal waters and a single genus, *Ceratium* (a Dinoflagellate) is by itself the subject matter of a monograph by Dr. R. Subrahmanyan (published by the Marine Biological Association of India); there are many species in the genus *Ceratium*. The members of the genus *Ceratium* are characterised by the presence of three horn-like projections at the three angles of the roughly triangular body (e.g. *C. tripos*). Another flagellate, *Noctiluca miliaris* is sometimes produced in very huge swarms in some localities in the sea spreading over several square miles of the surface water on some occasions, causing what is called the "Red Water Phenomenon" which is often associated with a large-scale mortality of fish and other creatures of the particular area in the sea (and also sometimes in the neighbouring areas). A flagellate species called *Protoeuglena noctilucae* lives inside the emptied "shells" of *Noctiluca,* the latter genus having no chlorophyll in the cells as an exception to the rule in the group. The presence of *P. noctilucae* can cause a green discolouration of the sea but there is no associated mortality of organisms. Still another species of flagellate called *Hornellia marina* is also associated with this green discolouration, again with no toxicity involved for the other organisms.

(b) The Porifera or Sponges

These are among the lowest evolved of the multicellular animals and in a way their appearance in nature makes one think of the encrusting and plant-like creatures and they sometimes even look like inanimate objects though mostly soft to touch. Sponges are predominantly marine and the familiar bath sponge is only a cured skeleton of a colony of sponges which are all porous. The body of a sponge is attached to a substratum, usually a rock or boulder or small stone; but the animal includes a freeswimming stage in its life history, called the *Parenchymula* larva which is provided with motile cilia for swimming in the seawater. In due course the larva settles down on a substratum, loses its cilia and the cells multiply gradually to result in the adult body which is a colony of several units provided with a skeleton of fibrous material strengthened by large numbers of what are called spicules which are hard pieces of variable shape and made of calcareous or siliceous material. There are many species of sponges and they fall into three main groups namely, the Calcarea, the Hexactinellida and the Demospongiae. The size, shape and chemical composition of the spicules are among the chief taxonomic characters used in the classification of the Porifera. The Calcarea have a skeleton made of calcium carbonate; the Hexactinellida have a skeleton of silica with spicules of six-rayed or Hexactinellid pattern, while the Demospongiae have skeletal elements (spicules) of silica or spongin fibres or both these, without the six-rayed pattern. There are numerous species of sponges in our seas. Thomas (1983) mentions 481of these in the Indian region between 60°E and 100° E down to 10° S latitude; there seem to be 275 species in the Gulf of Mannar, Palk bay and Sri Lanka waters. "Our knowledge of the marine sponges of India after Schulze's report on the Hexactinellida of the "Investigator" and Carter's papers on Tetraxinellida is due to Annandale (1914, 1915), Dendy (1926), Kumar (1925), Burton (1928 and 1932) and Burton and Rao (1932). Annanandale studied the Clionid sponges, common oysters, mussels and other molluscs and the sponges parasitic on them, and he gave an account of the biological differences of the sponge fauna north and south of the Palk straits connecting the Gulf of Mannar with the Bay of Bengal; Kumar, Burton and Rao advanced knowledge of our sponges in the Indian littoral, while Dendy and Burton made several interesting additions to our knowledge of the deep sea sponges. The interesting alga-like Euceratose sponge *Cryptospongia enigmatica* from a depth of 1045 fathoms and the peculiar spongid *Stenospongia aligera* from Mergui (65 fathoms) with chitinoid cuticle and the pores restricted to the undersurface of a number of lateral wing-like processes are among the forms described" (Rao, 1938).

The Gulf of Mannar and Palk bay have been well covered in the work on sponges in recent decades. Thomas (1968, 1970, 1970b etc) has contributed very significantly as the latest investigator in the area. Of the group Demospongiae itself he has recorded 125 species in this part of the sea, these falling into 84 genera and 33 families and including eight species considered new to science and 20 species as new records to the Indian region.

Burton (1937, 1959), Ali (1956 and 1956a) and Sivaramakrishnan (1951) have also added substantially to our knowledge of the sponge fauna in our area in recent decades earlier to Thomas, the last two authors working in Madras area.

(c) The Coelenterata

This phylum has a very large number of marine representatives, some of the smaller groups being entirely marine and unrepresented in the freshwaters. The phylum is defined as consisting of diploblastic or two-layered animals the body wall being made of two layers of cells namely, the outer ectoderm and the inner endoderm; the latter encloses a cavity which serves the functions of digestive and most other systems of higher animals. Both the ingestion of food and the egestion of the undigested remains of the ingested items takes place in this group through a single aperture called the "mouth" situated on an elevated part called the manubrium at the distal end of the attached body. The ectoderm layer of the body cells has a nervous function also and possesses some stinging cells called the nematocysts that can help in the defence and offence in the activities of the animal. There are sessile as well free-swimming animals in the Coelenterata and most of the species have both the stages in one life history, each alternating with the other stage. The sessile stage (attached to a substratum) reproduces by an asexual method of fission or budding, while the mobile stage reproduces by the sexual method that is, by the production of the spermatozoa (male cells) and ova (the female cells) which unite in the water and result in the embryo.

The phylum Coelenterata is divided into four main classes namely, Hydrozoa, Scyphozoa, Anthozoa and the Ctenophora the last group ranking as a separate Phylum these days, the first three classes then grouped together as a phylum *Cnidaria*. The Hydrozoans are relatively simple in body structure but the group includes several colonial forms and a few families form skeletons which resemble the corals, the true corals belonging to the Class Anthozoa in the same phylum. The Siphonophora exemplified by *Physalia* (The Portuguese man-of-war) and *Porpita* form a specialized group among the Hydrozoans and have received some attention in our studies also. The Scyphozoans include the familiar jellyfishes with umbrella-like bodies and have body-diameters ranging from a few millimetres to several feet; they swim with the help of the strong musculature present in the umbrella as well as in the projections from the lower aspect of the body called the "arms". *Aurelia* and *Acromitus* are among the commonest examples of jellyfishes which are actually the adults or sexual stages of these animals; their bodies have each a quadrilateral radial symmetry possessing four sets of gonads or sexual organs which dehisce their products when ripe into the common and single body cavity (the coelenteric cavity) and thence to the outside. After fertilization and segmentation an embryo is formed which further develops into a ciliated larval stage; this larva after some free swimming existence attaches itself onto a substratum, elongates itself and obtains a hydra-like appearance. This stage soon develops transverse constrictions which cut out the body into a number of circular discs which develop tentacles at the margin in a radially symmetrical arrangement. These discs detach themselves from the single parent body of the attached stage and become stages called the Ephyrae, these becoming the fore-runners of the future jellyfishes from which stage the life history originally started. A number of Ephyrae arising out of a single attached stage may remain together for sometime giving a chain-like appearance of Ephyrae with their tentacles projecting out from the margin of each.

The Anthozoa include the famous corals and the sea anemones. Though there are a few coral-forming species among the Hydrozoa also, they are distinct from the Anthozoan skeletons which are known as the true corals. The sea anemones are quite common among the inhabitants of the intertidal region and some species occur in the seabottom, mostly in the shallower areas. The anemones are sometimes attached to floating wood and similar vegetable materials also. The occurrence of the sea anemone (e.g. *Adamsia*) on the shells of the hermit crab leads to the commonly known case of commensalism in the littoral zone. The sea anemones are tubular-bodied animals attached to the substratum by one end of the elongated body and having the mouth at the distal end of the tube surrounded by a whorl of tentacles which possess nematocysts characteristic of the phylum. Some anemones are beautifully coloured especially on the tentacles, a pinkish brown colour being frequent in some species. They are rather delicate creatures and cannot withstand beyond some limits, any long exposure on the shore by the receding tides. They are hence most abundant only in the lower reaches of the shore near the low water levels. When exposed to sun and air during low tide the body which is normally found blooming out and large like a flower contracts itself into a bud-like structure with the tentacles and the oral cone well folded in. As soon as the tide rises again the body of the anemone easily and quickly expands and blooms out in response to the submergence in the water. The larval stages of the sea anemone have a freeswimming existence and these larvae attach themselves in due course to some substrata and grow into adult anemones. Thus both sexual and asexual methods are combined in the life history of the sea anemones, the former taking place by the male and female gametes uniting in the water into which they are released, and the latter by a process of budding.

An adaptive ability possessed by the sea anemones is the power of osmoregulation; though they are such delicate creatures it has been noticed on the west coast of India that some of them are among the few creatures that stand out as successful survivors in the sea when the southwest monsoon weather disturbs the seabottom during June-September every year, resulting in the formation of mudbanks and slush-flows that cause migration and destruction of many of the creatures of the sea-margin.

Unlike the sea anemones, the corals are mostly colonial creatures though there are a few solitary forms among these also. Each unit in a coral colony is a polype as in the Hydrozoa and has a tubular structure. All the individual units of the colony secrete calcareous skeletons (chitinous skeletons in some species) around themselves, the entire colony expanding and growing above the earlier-formed part of the skeletons from time to time. The skeleton thus developed is hard enough and provides spaces where the polypes and the connecting parts of the colony can remain fully protected and can carry out their life activities. The process of formation of the skeleton is naturally slow but there are literally "mountains" of corals thus formed over long periods of time in different part of the world, and some oceanic islands in the Pacific and the Indian Oceans are said to be of pure coral formation only. The Lakshadweep Islands of the Arabian Sea on the southwest coast of India are examples of extensive atoll formation these atolls being one of the three types of coral formation; these three types of coral formation are: (1) the fringing reefs, (2) the barrier reefs and (3) the atolls. As and when the living part of the coral colonies reach the upper levels by growing over the lower parts of hardened coral, the latter (the lower parts) seem to become gradually life-less rocks which still serve as substrata for other animals also along with the newly growing coral polypes. Some algae are known to grow and thrive in a symbiotic association with the corals in many situations.

The most extensive barrier reef of the world is the Great Barrier Reef of Australia with a reported length of about 1250 miles along the coastline of that country. But the fringing reefs are perhaps the most common of coral reefs and they are found on and near the shores in many tropical areas of the sea. The atolls are usually associated with many island formations in different parts of the tropical oceans. Many species of coral also occur along the Indian coasts such as off Rameswaram on the east coast in Tamil Nadu.

Common examples of solitary corals are *Fungia* and *Dendrophyllia* (fig. 32 & 31). Among the colonial corals the madreporarians have skeletons which are branching tree-like, the branches and stem being nearly but not often quite cylindrical, with polypes all round the stem and branches in cup-like depressions. The brain coral (*Coerulia*) has a grooved skeleton; *Tubipora* has a skeleton which consists of bundles of tubes arranged in a series of tiers, the tubes being all red in colour. Another coral of red colour is the precious red coral of commerce, *Corallium rubrum. Alcyonium* or "dead man's fingers" is an example of a chitinous skeleton, the branches of the coral being roughly finger-like and justifying this name. The sea-pen, *Pennatula* and the sea-fan, *Gorgonia,* are other common examples of the Alcyonarian corals.

In recent years a good deal has been added to our knowledge of corals and their biology in the Indian seas (Symposium on Corals and Coral Reefs, *Mar. Biol. Assoc. India,* 1972); Gopinatha Pillai (1972) and some others have played a significant part in these additions to knowledge. According to Thomas and Rani Mary George (1987) 22 species ot seafans (Gorgonacea) are being at present times exported from India, these belonging to 7 families and 15 genera. The Gorgonids acquired commercial importance because some useful chemicals such as prostaglandins and terpinoids of medicinal value are extracted from them. It appears that this importance was discovered very recently only (say, from 1969 onwards). However, we have still to wait and watch the extent of availability of the material as well as its utility and utilization in actual practice. Endean (*Encyclopaedia of Marine Resources,* 1969, p. 63) says: "Terpinoid substances possessing antibiotic activity have been isolated from extracts of gorgonians and alcyonarians but there is a strong possibility that these substances are produced by symbiotic dinoflagellates found in the tissues of gorgonians and alcyonarians."

Coral rocks are of great economic importance, being mainly useful in the cement industry and material for road-jelly in some places (for example in Tamilnadu near Rameswaram and other parts of the Ramanathapuram District). Figs. 31 & 32 show skeletons of two common corals.

The Ctenophora is a small group of Coelenterates now grouped as a phylum by itself, consisting of small freeswimming animals with a roughly globular body outline; there are several rows of vertically arranged cilia from one pole of the body to the other, these vertical rows simulating comb-like structures capable of a wavy swimming movement which helps the animal in its movements in the water. These animals are also called comb-jellies because of these comb-like rows of cilia, and also perhaps because of the mass of the body being soft and jelly-like as in the jelly-fishes. Common examples of the comb jellies are *Coeloplana, Ctenoplana* and *Beroe.*

(d) The Platyhelinthes

These are the flatworms; there are three main classes namely, the Turbellaria, the Trematoda and the Cestoda in the phylum. The Turbellaria are freeswimming ciliated forms while the trematodes and cestodes are parasitic; the former are mainly ectoparasites infesting the hosts from the outside while the latter are endoparasites, living inside the bodies of the host creatures. These parasitic forms are important in marine ecology because they can affect the fishes and other animals useful to mankind.

(e) The Nemathelminthes (Mainly Nematoda)

This old phylum includes what are popularly known as round worms or thread worms. Freeliving nematodes are plentiful in the sea particularly in the seabottom and in the intertidal region. Some nematodes are parasitic. Many forms still seem to be awaiting correct identification though a large number have already been studied from different regions.

(f) The Nemertinea

These are also unsegmented worms living as free animals in the seabottom and in many ecological situations on the seashore. The body is usually slightly flattened and elongated and always unsegmented. Pinkish and reddish forms of nemertines are quite common in the inshore trawl and dredge collections along the north Kerala coast. Many animals of this group also still remain to be identified with definiteness. Like many other groups of sea animals these too serve as food for other bigger animals especially the fishes which are most useful to mankind. The Nematodes, Nemertineans and some other smaller groups sometimes rank as independent Phyla.

(g) The Annelida (or Annulata)

These are segmented worms in which the elongated body is divisible into a number of serial parts or segments which are similar to one another in many structural features and also marked externally by constrictions or grooves encircling the body after each segment. In the group Polychaeta, each annulus or segment of the body also possesses on the outside, a pair of lateral muscular projections called parapodia; these parapodia or lateral feet as they are sometimes called, are locomotory organs possessing also bundles of stiff bristles embedded in them and called setae. Parapodia are absent in earthworms and in the leeches or Hirudinea.

The Annelids are among the highly organised triploblastic metazoans having three well-differentiated layers of cell-types in the body structure . These three layers can be most clearly seen close together in the embryo and are called the endoderm (or innermost layer of cells), ectoderm (or the outermost layer of cells) and the mesoderm (or the layer of cells placed intermediate between the above two layers) the last mentioned layer is absent in the more primitive Metazoa like the Coelenterata. In the later development of the embryos of all metazoans the ectoderm gives rise to outer epithelia and the nervous system. In these and all the Chordata the endoderm gives rise to the digestive system as well as skeletal system, the coelomic cavities and the tissues inside them (and forming their boundaries) being formed by the mesoderm. In the earliest embryo, as is well known what is seen is only a mass of dividing cells which, in due course, arrange themselves into one layer and then two layers namely, the ectoderm and the endoderm usually by a process of invagination; the single-layered blastula thus becomes a two-layered gastrula enclosing the gastrocoele cavity inside it; this "gastric" cavity is surrounded immediately by the endoderm and covering this on the outside is the ectoderm; from the endoderm later arise the enteric canal and the various glandular and other tissues associated with the same. The connective tissue, muscles and the reproductory and excretory organs are all developed out of the mesoderm. The coelome or body cavity is a characteristic feature of all advanced metazoans, and this cavity contains within itself the various organs and structures of the general viscera. In the coelenterates there is only a single cavity in the body as stated earlier, with the ectoderm and endoderm enclosing it and remaining close together. This single cavity is called the coelenteron or coelenteric cavity within which digestion and all the other life functions take place, the undigested "debris" being ejected through the "mouth" opening itself. Unlike the coelenterates and the sponges, the Annelids possess not only a coelomic cavity but also a separate enteron or alimentary canal; in addition there are well-formed circulatory, nervous and excretory systems and atleast a partial metameric segmentation is noticeable in all of these.

The alimentary canal has an anterior opening, the mouth, and also a posterior opening, the anus, through which the undigested matter is excreted. The nervous system in the annelids consists of a fused pair of thickenings or ganglia together called the "brain" at the anterior end of the body; there is a long double nerve cord running along the body dorsal to the alimentary canal; nerves are given off to the head region from the cerebral ganglion, the brain. The excretory system consists of special sacs called *nephridia* which are paired and are located in some segments of the body. The circulatory system consists of elongated dorsal and ventral blood vessels provided with red blood in some species. The two long blood vessels are connected by lateral loops in some segments and these loops, slightly swollen with blood, are called the lateral hearts. Thus the Annelids are rather highly evolved animals with a very great deal of specialisation and differentiation in all the systems as compared to the earlier animal groups (so far considered).

Annelids are well-represented in the sea and the order most represented seems to be Polychaeta of the class Chaetopoda. Polychaetes are commonly represented by the lugworm, *Nereis* (Figure. 56) several species of which occur in the shore and shallow sea in the various areas of the world. The larvae and early juveniles may occur in the plankton at the surface, midwater or at the bottom in practically any part of the sea during certain seasons of the year; the adult full-grown polychaetes are mostly at the bottom of the sea though in a few cases these may continue a planktonic existence itself. The first free larval stage of the polychaetes is a top-shaped ciliated creature called the *Trochophore* with circlets and several regular rows of the cilia which help in locomotion; in addition there is a tuft of longer cilia on the broad-end of the top-shaped body (Figure 18 & 19). *Prionospio pinnata* is a very common polychaete in the inshore sea of the west coast of India, and it occurs in the form of its larval stages in the surface plankton, its metamorphosing and early juvenile stages in the midwater layers and in the adult stage at the seabottom. This worm is important as fish food and has been found to be a favourite item in the food of the Malabar sole, *Cynoglossus semifasciatus* which is an economically important food fish of the north Kerala and south Karnataka coasts.

Polychaetes are among common sea animals that have received some attention for experimental work in some of the Indian marine laboratories; among these must be mentioned the early work on osmoregulation by Krisnamoorthi (1951 etc) on *Marphysa gravelyi.*

(h) Pogonophora

An interesting group of animals has to be mentioned before passing on to the higher phyla of the animal kingdom; this is the *Pogonophora.* Several species of this group have recently come to light, the indian waters also being not left out in this respect. A peculiarity of the Pogonophores is that they do not have an alimentary canal though they appear to be higher in the scale of evolution than the groups of worms *sensu strictu,* namely the Platyhelminthes and the Nemathelminthes. They have a tubiculous habit, the chitinous transparent tubes in which they live being secreted by the animals themselves during their growth from the early stages. Some workers treated them as protochordates but this view is not supported by others.

(i) Kinorhyncha

Another group to be mentioned here is the *Kinorhyncha,* which is often treated as an appendix to the Annelida and sometimes even as an independent phylum, but mostly as a class of the Nemathelminthes or as an appendix to Annelida. These are tiny creatures not more than a centimetre in length and have a worm-like appearance and also an annulated body. The most important and interesting feature about these animals however, is their habitat; *they live in the interface between the surface layer of the mud at the bottom of the sea and the overlying seawater.* Their occurrence is thus restricted to such a narrow stratum in the sea that specially designed instruments alone can capture them as they are not caught either in the Grab and Dredge used for the bottom fauna collections, or in the plankton nets used for the collection of the floating organisms of the water at the different depths. Dr. Higgins of the United States of America who visited this country in connection with the International Indian Ocean Expedition in the late sixties, seems to have been the first to construct a special instrument and to demonstrate a collection of these animals in our country so far.

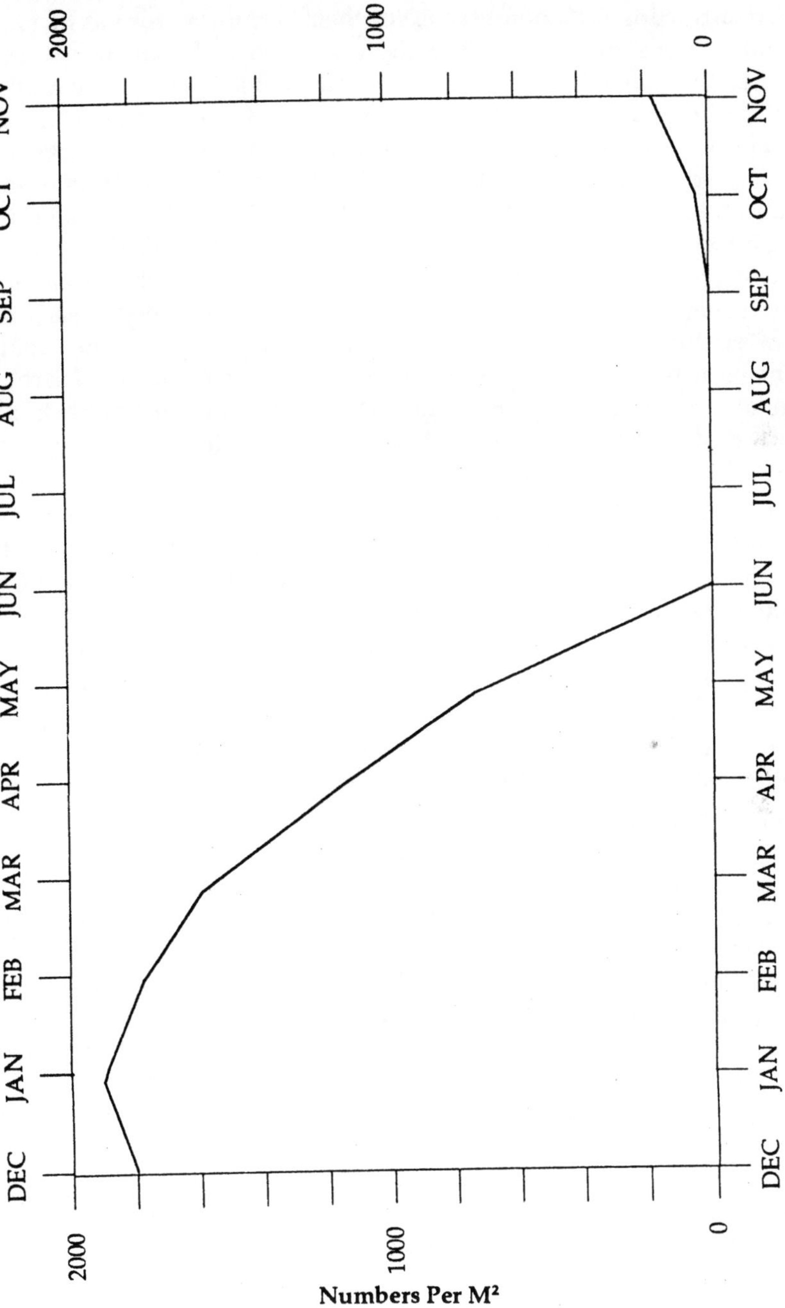

Fig. 36: Numerical Abundance of *Prionospio pinnata* in the inshore sea at West Hill, Calicut in 1949-50 at 0-10F. (Data from Seshappa, 1953).

(j) The Arthropoda

This is a very large phylum including in it the class Insecta which comprises economically the most important animals of the Animal Kingdom, most of the insects however attracting attention because of their harmful nature to mankind and to the materials useful to man; but the class also includes many useful creatures such as the honey bee, the silkworm and the lac insect. The animals coming under Arthropoda also have a metameric segmentation of the body and possess generally a higher level of organisation both in structure and function. A most noteworthy specialisation noticed is the formation of a head region (and even a very distinct head) as distinct from the rest of the body, this specialisation reaching its perfection in the Insecta. The process of cephalisation or head differentiation may perhaps be said to have started in the Annelida in a small way but there it is at best only an indication of the beginning of cephalisation in the process of evolution from the lower to the higher groups of the Animal Kingdom. The occurrence of a clear neck region is seen in the insects. There is a concentration of the sensory organs in the head in which the mouth is also situated with well-developed jaws or jaw-like appendages. Insects are few in the sea, restricted to the bottom.

The outer surface of the body in the arthropods is protected in a hard covering of a chitinous substance in each segment or group of segments. It is actually only the abdominal segments that retain their separation, those of the anterior region of the body being fused together.

The commonest examples of arthropods from the sea are the crabs and prawns (different species of which occur in freshwaters also), and lobsters and seaspiders. The creatures of the last category mentioned belong to the group *Pycnogonida* which are not spiders at all but certain specialised creatures appearing somewhat like spiders and inhabiting the seabottom.

The phylum Arthropoda is divided into five main classes: the Crustacea, the Onychophora, the Myriapoda, the Insecta and the Arachnida. The Crustacea is a large group and includes the prawns, crabs and lobsters along with several other forms. All of these have characteristic chitinous shell on the body, possess metamerically arranged appendages on the different segments of the body and breathe by means of structures called the gills which are branched feather-like and are filamentous outgrowths from certain parts of the body; they are provided with fine capillaries and are well-suited for the function of respiration. The appendages of the different regions of the body are modified for different functions according to their positions; thus, some of these on the head are sensory feelers while some others are jaw-like structures helping in the capture and mastication of the food items; still others are used for locomotion (in the thoracic region) being modified into legs, while those in the abdominal region help in swimming and are called the swimmerets or pleopods.

The eyes of arthropods are of a compound and mosaic type, each eye being made up actually by a group of many distinct units of simple eyes each of these providing a visual picture of only a part of the object seen, but a full picture of the entire object in view being available only as a result of the integration of all the images formed by the various part-pictures which are automatically juxtaposed in a successive order.

From the economic point of view the prawns and lobsters are among the most important forms from the sea, especially in India at present, as large quantities of prawns are exported to many foreign countries. The production of prawns seems to be steadily increasing in recent years—though occasional set-backs are not altogether unknown—and increasing effort is being applied in the country for a higher production and exploitation of this resource. Many organisations—under the Government of India and the various State Governmental organisations in addition to private enterprise—have taken up activities related to prawn culture with encouraging results.

Lobsters are at present being caught in relatively small quantities only but these forms are likely to assume more and more importance with culture experiments also where possible. At present the lobster "tails" exported are in rather small quantities but weight for weight they fetch more money in the trade than most other items of sea food.

Among the commonest marine prawns are *Metapenaeus dobsoni, M. affinis, M. monoceros, Parapenaeopsis stylifera* and *Penaeus indicus. Macrobrachium rosenbergi* the large backwater prawn called "*Konchu*" (in Kerala) and other species of *Macrobrachium* appear to be giving good results in culture experiments also; on these, more will be mentioned later under *Marine Fisheries*.

The lobsters are considered a delicacy both by Indian and foreign consumers and hence fetch higher prices than prawns. *Panilurus homarus* is a common lobster occurring at many places. The movements of the lobsters in the sea form an interesting and useful subject for study and some experiments have been conducted by the Central Marine Fisheries Research Institute of Cochin at different places for following the paths of their migrations along the coast by means of marking and tagging.

The commercial prawns and crabs of our country may be said to be occurring mostly in the estuaries and the shallow inshore areas of the sea. Some estuarine prawns have been studied for their osmoregulatory ability which enables them to survive successfully in the environment of changing salinities. Juveniles of many species of prawns seem to be abundant only in the estuaries though the adults may occur in both the sea and the estuary. However, there are some species which form a fishery only in the estuaries while there are others which are exclusively or particularly abundant only in the sea.

Among other interesting examples of marine crustaceans may be mentioned the barnacles (e.g. *Balanus*), copepods, amphipods and isopods. The copepods are among the most important constituents of the plankton in the sea everywhere and they form the very common food item for many fishes. Several species of isopods and amphipods occur in the benthos of the estuarine as well as the shallow sea environments, and these do often occur in the plankton also. The plankton at times abounds in various larval stages of many kinds of crustaceans including also the barnacles which in the adult stage are firmly fixed to rocks.

The common rock barnacle belongs to the species *Balanus balanoides;* barnacles were among the interesting creatures studied by the famous Charles Darwin also.

While there are many economically useful forms among the arthropods there are several harmful forms also and the animals of this group are thus important for study from both considerations. As stated earlier the entire group *Insecta* is of high importance economically and includes in itself the highest number of species among the different classes of Arthropods and perhaps among the different *Metazoan* phyla also. Among the harmful Arthropods that are important may also be mentioned the wood-boring isopod *Limnoria lignorum* and the fouling cirripedes.

The pycnogonids or sea spiders are of considerable academic interest because of their morphology and their place in evolution.

(k) The Mollusca

The animals of this phylum are mostly provided with a much harder protective shell over the body than the arthropods, the shell here being made of hard calcareous matter unlike that of the arthropods where the shell is chitinous and relatively less hard. The metameric segmentation noticed in the annelids and arthropods is absent here. The body is bilaterally symmetrical and there is a complete alimentary canal with accessary glandular structures. The appendages such as parapodia of the annelids and the walking legs and swimmerets of the arthropods have also disappeared. Locomotion takes place in the adult molluscs by means of a muscular organ called the *foot* which assumes different shapes and structures in the different classes of the phylum.

The phylum Mollusca is divided into five main classes depending on the nature of the pedal organ: the Amphineura, Scaphopoda, Lamellibranchiata, Gastropoda and the Cephalopoda. The Amphineura are the *Chitons* which live attached to rocks protected in convexly flattened shells (like limpet shells with a little more flattening), the shell in each individual consisting of some transverse pieces held together by musculature. The chitons are able to move a little more easily than the limpets which have a shell of a single piece only and have extremely limited movements, the shell of the chitons having the several pieces held together by muscles which can be utilised in the movement of the entire animal to make locomotion apparently easier for the creature. Nevertheless the chitons also move about much less than the gastropods which form the next class (the snails) and which normally have quite frequent crawling movements on the substratum.

The Gastropods are "stomach-footed" animals with the body constructed in such a way that the visceral mass is carried over above the foot; when the animal moves about, the entire spirally coiled shell with the soft parts of the animal inside it is carried about above the pedal organ which crawls behind resembling a "tail". The snails are creatures of this group, their shells having a plate-like operculum covering and protecting the opening through which the head and foot can come out or go in according to necessity; the entire animal can remain hidden inside the shell and fully sheltered against any outside interference.

The Lamellibranchs have a laterally compressed two-valved shell, the parts of the shell being hinged together in the middle dorsally. The size, colour and sculpturing on the shell are variable from species to species. The foot of this group (which is also sometimes called the *Bivalvia*) is plough-share- shaped and is situated in the anteroventral aspect of the body with a laterally compressed and a bluntly pointed anterior end. On either side beneath the shell between it and the body proper is a membrane called the mantle which encloses a mantle cavity in which other soft parts of the animal are also contained. The edges of the mantle of the two sides can be closely applied together when necessary and separated at other times.

The Cephalopoda are unlike all the above groups of Mollusca, what may be called the "head-footed mollusca" with a foot or pedal organ so modified that it surrounds the mouth in the form of a number of long thick muscular processes called "arms" useful in locomotion as well as in offence and defence and feeding activity.

Molluscs occur both in the sea and in the fresh water, but some are entirely marine. A common example of the Lamellibranchiata (also called Pelecypoda) is the edible green mussel, *Perna indica* (which was formerly known as *Mytilus viridis*). *Modiolus* is another bivalve genus of the same group but it is a tiny form compared to the mussel and frequents estuarine and marine intertidal environments. *Meretrix* is a genus including the common backwater clam and other species and varieties. The edible oyster *Ostrea madrasensis* and rock oyster *Ostrea cucullata* are also members of the class Pelecypoda. The food of all these organisms consists of minute creatures occurring in the water moving about inside the mantle cavity. *Martesia striata* and *Pholas orientalis* are common species boring into substrata and belonging to the same class.

Some molluscs are very harmful to wood; a very common example of this kind is *Teredo navalis*, the ship worm which is a specially well-adapted animal for this type of living by eating through the wood of ships by the help of a chemical secretion and living inside the burrow thus formed inside the wood. The modern ship-builders seem to have switched over largely to steel (wherever possible) as building material for the vessels as this is fully proof against this worm-like animal which used to cause great havoc and damage to ships constructed entirely with wood as the basic material. The bivalve shells of *Teredo navalis* are so highly reduced and readily unrecognisable that the animal can be confused for a worm at first sight. However it is also a bivalve mollusc like other Pelecypoda with the general appearance of the group being masked in the adult animal.

Some bivalves are known to possess some degree of ability to osmoregulate; the estuarine species of Mytilus form examples of this. Mussels, clams and oysters are all sometimes reported to suffer some degree of mortality owing to factory effluents polluting the habitat occupied by them in nature.

Vekataraman (1955) and others have done some work on this aspect in India as in the western countries but still a lot of work seems to be due.

The Pearl Oyster, *Pinctada fucata* (=*P. margaretifera*) grows as a benthic inhabitant in the shallow sea off Tuticorin and a few other localities along the Indian coast. There is some traditional pearl fishery off the Kathiawar coast on the northwest coast of India and also in the Gulf of Mannar off the coasts of Sri Lanka and India, Tuticorin being included in the latter. Orton and Amrithalingam have done some pioneering work on the pearl oyster form the Sri Lanka side in the past and in recent years much progress has been made in India by way of experimental pearl culture work (Qasim and Alagarswamy).

The pearl is a secretion of a glossy substance around a tiny foreign particle that gets into the space between the mantle and the shell of the animal in the natural habitat and general conditions of environment. Artificial pearl formation can be induced by the introduction of minute sand or other particles between the mantle and the shell of the living animal. The pearl is secreted by the living animal from the mantle epithelium in response to the irritation caused by the foreign particle. Japan has been the pioneer in developing and commercially exploiting the cultured and artificial pearl industry, the name of Dr. Mikimoto of that country standing out as the individual first pioneer in this line of investigation and work.

The Pelecypods occur in the deep seas also though perhaps they are relatively small in numbers there when compared to the near-shore and shallower water areas of the sea. The wedge clam *Donax cuneatus* is a bivalve of the intertidal region but confined mostly to the sands of the low water level and thereabout, and extending to a short distance into the subtidal zone also occasionally when there is fine sand available there. This animal is useful as food for other large animals and can also be used as bait to catch fishes, in addition to its being useful as man's food also like the mussel and oyster though less popular.

Among the gastropods the top shells of the genus *Trochus* occur in large quantities around the Andaman islands and have been subject to a traditional fishery for more than a century. These shells can be used for making buttons and the meat can be used as bait. Fig. 37 shows the developmental stages of the snail *Littorina saxatilis ruddissima* (Bean).

The gastropod foot is divisible into three parts namely, an anterior part carrying the head with its tentacles, eyes and the mouth, a middle part over which the entire viscera or mass of the animal rests, and a posterior part which trails behind like a "tail" for the animal. On the posterior part stated, there is a plate-like structure called the operculum which can be closed over the opening of the shell when the entire animal withdraws inside and needs protection. The operculum is of great help particularly to the intertidal gastropods as a means of protection not only against predators but also against the desiccation during low tides. Another special structure possessed by the gastropods is a rasping organ called the radula which is present in the oesophagus or gullet of the animal. During feeding the radula is brought outside and used for masticating the food material, and when not in use, the organ can be spirally rolled inside and kept at rest till called for again for work. Many snails have been found to be vegetarian and others omnivorous, the latter being specially the case with the larger snails. The relative growth of the radula along with its size and shape as well as the numbers and arrangement of the rasping teeth on it are of taxonomic importance in the different forms of snails.

The common cowry *Cypraea moneta* is an example of marine gastropods well known to most people; cowries are known to have been used as a sort of coinage in ancient times, and they are at present very useful in making various ornaments and children's playthings. Some tribal people such as the Lambanis of the south decorate themselves profusely with garlands and bangles made of small polished and variously coloured cowries, the shining white cowries being liked most, perhaps. Some gastropods have shells of various beautiful colours, colour patterns and shapes. The flesh of many gastropods is edible. The edible abalone (*Haliotis*) forms a fishery of great importance in the Australian seas. The limpets are gastropods with the shell vertically compressed or flattened a great deal, the spiral appearance vanishing completely in an extreme modification.

Common living examples of the Cephalopoda are *Sepia, Loligo*, and *Octopus*; the pearly shelled *Nautilus* also belongs to the Cephalopoda. *Loligo* and *Sepia* are of relatively small size but *Octopus* is the largest representative of the class. The foot is modified in this group into eight or ten long muscular projections called the arms surrounding the mouth; behind this part there is a tubular structure called the exhalent siphon through which the animal can throw out the rejected wastes and fluids from the body to the outside, that is, from the mantle cavity of the animal which is more compact and more enclosed when compared to the mantle cavity of the Lamelibranchs. The shell is often internal in the Cephalopoda and may be highly reduced and even practically absent. The arms have cup-like suckers on them; these suckers are very well developed and strong in forms such as the *Octopus* or the devil fish which is notorious for its fierce attacks on fishermen occasionally in the seas; fishermen are said to have had taken recourse to cutting away of the arms (of the devilfish) to free themselves from the suctorial attacks. The very large individuals of the *Octopus* seem to occur usually only in the oceanic regions, but the animal is capable of living in the surface waters as well as at middle depths. Smaller devilfishes are seen in the inshore seas also.

Sepia and *Loligo* have ten oral arms and belong to the Decapoda while *Octopus* has only eight arms and belongs to the octopoda. Both these forms form fisheries of some importance in several parts of our coast, both being edible and also useful as bait for capturing the true fishes. *Sepia* has the ability to eject a dark ink-like fluid from its exhalent siphon on irritation (on any disturbance) and to escape under cover of the dark colour.

(l) The Echinodermata

The animals coming under this phylum seem to be exclusively marine in their habitat. As the term Echinodermata indicates, these are all spiny-skinned

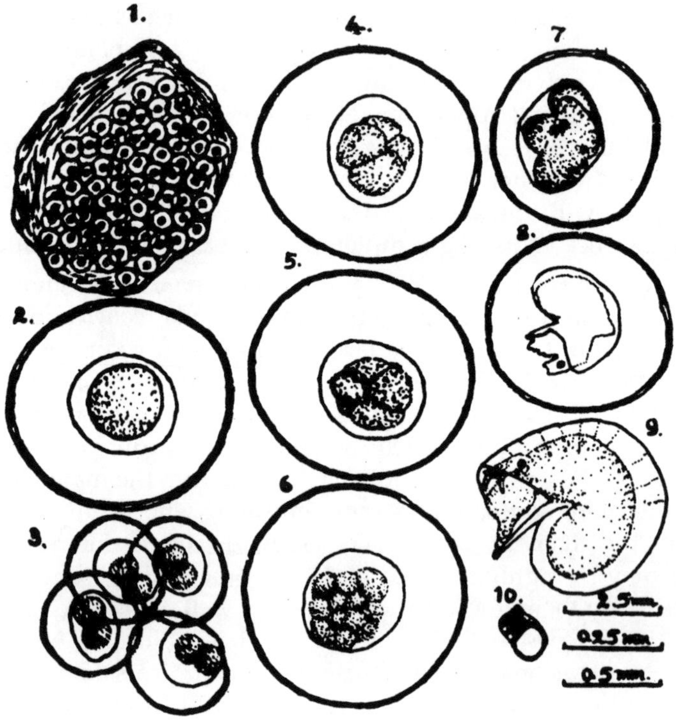

Fig. 37. Segmentation and developmental stages in *Littorina saxatilis rudissima* (Bean): 1.egg mass; 2. unsegmented egg; 3. 2-celled stage; 4. 4-celled stage; 5. 8-celled stage; 6. 12-celled stage; 7. & 8. two stages of shelled Veliger still inside the capsule; 9. first crawling stage just hatched out; 10. early juvenile (re-drawn from Seshappa, 1976)

creatures. These Echinoderms have a radial symmetry in the construction and organisation of their bodies, as against the bilateral symmetry of the Mollusca and the other groups. The body in this phylum can be divided into similar equal parts in a number of its diameters. The majority of the Echinodermata are inhabitants of the shore and shallow sea but there are many deep sea forms also such as the sea lilies which belong to the class Crinoidea.

Five classes are recognised in the Phylum Echinodermata: these are: the Asteroidea (starfishes) (Fig. 28), the Ophiuroidea (brittle stars) (Fig. 29), the Echinoidea (sea urchins), the Holothuroidea (sea cucumbers), and the Crinoidea (the sea lilies).

The Crinoids are stalked animals while the others are all free-living forms. The Asteroidea are the starfishes commonly found on many stony and shingly shores. The body of a starfish (Fig.28) is star-shaped with usually five projections (from a central disc) called the arms which are broad at their attachment to the disc and taper towards the tip. There are a few species in which the number of the arms is six, seven or even more (upto thirteen in one particular form). A lost arm can often be regenerated almost to the extent of the original size and shape, and mutilated specimens with injured arms repaired fully are occasionally seen. On the ventral centre of the disc is the mouth from the five corners of which radiate grooves along the middle of each arm, called the ambulacral grooves. On either side of the ambulacral groove in each arm are rows of tubular projections with sucker-like ends and called the tube-feet. These are the locomotor organs of these animals and they can also help in the capture and subjugation of food creatures such as mussels and snails. Both the disc and the arms of the starfish are supported by a skeleton of calcareous plates which are somewhat flexibly attached to one another; there are also tubercles of various sizes on some of these plates, both on the arms and on the disc, dorsally, ventrolaterally and laterally. Mounted moveably on some of the tubercles on the oral side of the disc and along the sides of the ambulacral grooves of the arms are some spines which serve the purpose of protection to the contents of the grooves. The starfish lives with its oral surface applied to the substratum the anal aperture being situated a little eccentrically on the opposite side which is the aboral surface of the body. A system of tubes forming a network inside the body and in which the sea water is circulating is a speciality of the Echinodermata and it is called the water vascular system which has an external communication by an aperture on the aboral surface. *Asterias* and *Astropecten* are common examples of the world genera of starfishes.

The brittle stars (Fig. 29) of the class Ophiuroidea are similar to the starfishes in having a disc with usually five arms radiating outward from the disc. The arms of the brittle stars can bend and twist themselves a great deal but their skeleton is rather brittle and the arms snap off too easily when the live specimens are handled carelessly. The arms are also solid and do not show the ambulacral grooves as in the starfishes though the zone corresponding to the ambulacral groove is recognisable easily here also. The disc of the brittle stars is rounded and the arms start abruptly from the disc and are uniformly thick for the most part (though they do taper off at the ends) while in the starfishes the arms are thick near the disc and taper off distally and the appearance is such that the arms and the disc gradually merge into each other. The ophiuroid arm is also relatively very long and the appellation given to the group as Ophiuroidea seems to be apt because of the length, slender form and the coiling nature of these arms. The brittle stars are however, only the more tender and fragile and brittle relatives of the Asteroids; they also possess the tube feet and the spines and also the various internal systems as in the starfishes; feeding and digestion also take place similarly in both the groups.

The class Echinoidea includes the sea urchins which are literally Echinoderms (=spine-skinned forms) because the body surface in these animals is covered by a very large number of spines all over. This class has lost the starshaped body of the Starfishes and brittle stars, and on the other hand, the body here is of a roughly rounded shape with a roughly broadened oral aspect and a relatively rounded off aboral or anal aspect. The radial symmetry is clearly visible from the outside of the body which is enclosed in a firm shell of immovably attached (or even "fused") skeletal plates, whose individual outlines can however be clearly recognized. There are five ambulacral areas on the shell alternating with five inter-ambulacral areas both of which can be recognised in the dry shells by means of the double rows of tiny apertures on either side of each ambulacral area; these are the apertures for the passage of the tube feet of the same function and similar structure as in the two earlier classes though here the tube feet are longer and more slender. The calcareous plates of the shell have tubercles of various sizes and mounted on them by ball and socket joints are a large number of spines which taper off to a point at their extremities. Both the tube feet and the spines help in the locomotion of the animal as also in offence and defence. Another interesting structure found in the sea urchins and not found elsewhere in the phylum is a masticating organ called the Aristotle's Lantern; this is a jaw-apparatus made of a number of pieces of bone-like hard material held together by strong muscles and capable of use for mastication of food organisms. The Class Echinoidea also includes the flat-shelled cake urchins and the heart-shaped heart urchins both of which however, are not very common forms. While most Echinoids live on the surface of the seabottom, some live in burrows in the sandy shores. *Salmacis bicolor* is a common Indian sea urchin and is used for class work in colleges in different parts of the country. Gopala Aiyar (1966) has written a most valuable monograph on this form, this monograph serving as the guide book for the class work. Fig. 22 shows the *Echinopluteus* larva of the sea urchin.

The members of the Holothuroidea are popularly known as the sea cucumbers because of their elongated and cucumber-shaped body. If a sea urchin could lose its shell and have a soft leathery skin instead, and if the body could be imagined to be pulled long along the oral-aboral axis, the result would be roughly like the sea cucumber of the Holothurian class. But consequent on the elongation of the body as stated, the animal now does not rest on the substratum on its oral aspect but lies on one of its sides much like any of the worms of say the Annulata, with the mouth at one end of the prostrate body and the anus at the other end. The mouth of the sea cucumbers is surrounded by prominently marked tentacles. There is a slight modification of detail in the radial symmetry of the body in that the tube feet are well developed only along the radii on which the animal lies on the substratum while the tube feet are reduced or even absent on the other radii or ambulacral areas which are away from the substratum. The skin is thick and leathery and has calcareous spicules embedded in it forming a supporting skeleton; these spicules are what are left here of the shelled skeleton of the other classes of the phylum. All the other systems of the Echinodermata including also the water vascular system are present in the Holothuroidea also and in addition there are a pair of branching tubular structures from the cloaca into the body cavity and these are called respiratory trees because of the tree-like branching and their "respiratory" function. There are also peculiar long finger-like processes from the cloacal wall called the Cuveirian organs; these lobed structures can be protruded outside through the anus and abandoned by autotomy or self-mutilation when the animal is attacked by enemies. The Cuveirian organ can be completely regenerated after the autotomy; this peculiar structure (which is not found in any other group of animals) is often attractively coloured so that it serves as a snare to prospective enemies which, attracted by the colour may catch the Cuveirian organ which the sea cucumber can autotomise and escape unhurt except for temporary loss of that organ itself.

The sea cucumbers are the material for the "beche-de-mer" industry of the USSR, and these animals are used as food elsewhere in the world also. Not fully exploited as yet, this is a resource with future prospects for research and development, though at the moment we still do not seem to have a reliable picture of the extent of its availability in our seas.

The Crinoidea are the last group of the Echinodermata to be considered before we pass on to the next phylum of the animal Kingdom. These animals are also called the sea lilies in popular language and are again considerably different from the other four classes of Echinodermata while showing all the same, all the chief characters of the phylum justifying their inclusion in the same. They are essentially animals of the deeper waters of the sea and are always attached to the substratum by means of a stalk. The body shape of the sea lilies can be roughly compared to that of an inverted starfish with the aboral end drawn out into a stalk for attachment and with the arms branching out several times dichotomously. Consequent on this kind of modification in body structure both the oral and the anal apertures are situated on the free end of the disc only. The skeletal plates are present on the stalk and also on the arms and the disc. The crinoids include animals growing several feet in length and also smallbodied forms. Two examples of Indian sea lilies are: *Tropiometra,* and *Bathycrinus,* the former occurring in shallow waters and the latter in the deep sea off the bay of Bengal.

Apart from the Holothurians the other animals of the phylum Echinodermata do not seem to be of much economic importance from the point of view of human food value though of course most creatures can serve as food for the fishes and other useful animals of the sea in some form or the other. Skeletons of the starfishes and sea urchins have some ornamental value and can serve as paper weights and can be used for various kinds of fancy articles and toys. The larval stages of all the five classes of Echinodermata can "enter" the stomachs of fishes and other animals and thus serve as their food.

(m) The Chordata

(i) *Definition*–Protochordates: All the animals described so far are without a backbone (vertebral column) and also without its rudiment called notochord. The animals of the present phylum, the *Chordata* possess in one stage or the other of their lives, a cartilaginous skeletal rod along the back known as the notochord, developed as an offshoot from the enteric canal (which is developed from the embryonic endoderm). The chordates are all more highly evolved animals than the earlier phyla described here. The more primitive of the chordates are called the Protochordates which are exemplified by *Balanoglossus* (*Hemichordata*, Class *Enteropneusta*), the Ascidians (*Urochordata*, Class Tunicata) and *Amphioxus* (*Cephalochordata*). In these animals when the notochord is present it is not split up into segments and there is no bone formation. The notochord is present in the adults in *Amphioxus* but occurs only in the embryonic stages in some others, and it may be in some forms represented only by a small and insignificant rudiment. In the more advanced chordates the notochord leads to the formation of a vertebral column in later development, this being at first cartilaginous (remaining like that permanently in some forms), and bony later, the cartilaginous structures of the skeleton getting ossified or converted into bone as the animals grow up.

(ii) *Hemichordata:* (*Balanoglossus*): All the Protochordates are creatures of the sea. The Balanoglossids are wormlike animals with a localised distribution and the Krusadai Islands of Tamilnadu are among the few places in India where these are abundant; the common Balanoglossid species in these islands is *Ptychodera flava* which occurs buried in fine sand and mud of the low water mudflats. Figs. 20 & 21 show the *Tornaria* larva of *Balanoglossus*. Owing to its relative scarcity in our country, the animal is protected against unlimited destruction by a control over the numbers that may be collected by the interested workers (mostly Zoologists of various Universities) from time to time. Though not of any direct economic importance the Balanoglossids are of some importance from a scientific point of view, being a link between the higher Chordata and the more primitive phyla of the non-Chordate groups of the Animal Kingdom.

(iii) *Urochordata* : (Ascidians): The Tunicates or Ascidians have small sac-like bodies and many live attached to some substrata by one end of the body; in each individual there are two tubular openings at the distal end (separated by a small distance) namely, the exhalent and inhalent siphons; through these siphons water is taken in (inhalent siphon) and ejected out (exhalent siphon) after circulation inside. The two siphons are also situated at different levels. The exhalent siphon discharges the fluids from a cavity called the atrial cavity which surrounds the viscera and is comparable in function roughly to the mantle cavity of the molluscs; the other cavity of the body is the alimentary canal into which water enters through the inhalent siphon; the circulation of the fluids inside the atrial cavity takes place by the action of cilia present on the epithelium in the different regions; there is also the ability of the animal to contract and expand by muscular action to a limited extent. The ascidians (Fig. 35) are also called the sea squirts. Some ascidians are solitary and there are others which are colonial, there being modifications of body form and structure in both types. Free swimming species of (Pelagic) tunicates also exist less numerously. Two examples of Indian ascidians are *Herdmania* and *Ecteinascidia*. S.M. Das, a pioneer worker on Indian ascidians has written a monograph on *Herdmania (H. pallida)* which has been in use as a text book for advanced classes in Zoology for over half a century (Das 1936).

(iv) *Cephalochordata* : (*Amphioxus*): *Amphioxus* (or *Branchiostoma lanceolatus)* lives by burrowing in fine sand and can also swim by means of fins or fin-like structures. This animal is spindle-shaped and is only a few centimetres in length; it is semitransparent in the fresh state. This animal possesses a well-developed notochord which extends almost to its entire length except at the extreme ends (especially the anterior). There is a funnel-shaped cavity at the anterior end of the body and inside this funnel is the oral aperture of the animal, the funnel itself being surrounded by a circlet of small tentacles. The mouth leads into the next part of the alimentary canal known as the oesophagus and then to the pharynx which has a series of slits in its walls on either side, the water entering by the mouth, and leaving the pharynx through these slits known as the gill slits; respiration takes place during the passage of the current of water through the gill slits. The food particles are also supplied through the mouth by this incoming water. The undigested remains of the food particles are ejected through the anal aperture situated at the posterior end of the body subterminally on the ventral aspect. There is a longitudinal nerve cord along the body lying along and above the notochord which is itself above the alimentary canal. The anterior end of the nerve cord is enlarged into a primitive brain.

(v) *Vertebrata*: (a) **Agnatha**-(Cyclostomata): Among the vertebrate Chordates all except a small number of primitive forms are provided with definite jaws as found in the higher animals that we see commonly all over the world. There is however, a small group of jawless animals at the lower end of the Vertebrate evolutionary tree, these animals being known as the Agnatha or Cyclostomata, and all the remaining groups of animals with jaws among the Vertebrata being together called the Gnathostomata. The word Cyclostomata means round-mouthed animals and these are exemplified by the hagfishes and the lampreys. The lampreys have representatives in both the sea and rivers while the hagfishes seem to be exclusively marine. The bodies of these round-mouthed fishes are elongated, eel-like and cylindrical; there is a round suctorial oral apparatus at the anterior end, this consisting of a funnel in the centre of which is the actual mouth surrounded by sets of rasping teeth. This animal can apply itself to the surface of the body of bigger animals by means of the oral apparatus and tear off and swallow the flesh and blood of such victims. Cyclostomes have no paired fins; there is an unpaired nostril situated dorsally on the head and there are seven pairs of gill slits behind the eyes, one set on either side. The tail of the lampreys is flattened and helps in the swimming movements; their skin is soft and covered with a sticky slime secreted by itself and there are no scales. The lateral line sensory organs do not possess the sensory canals noticed in the true fishes of Gnathostomata.

Sexes are separate and the sexual products are dehisced into the surrounding water where fertilisation of the ova takes place. The fertilised egg leads after segmentation and early development into a larval stage known as the *Ammocetes* larva. This larva has an elongated body, a median fin fold and a semi-circular upper lip; there are no teeth; the eyes are delicate and covered by skin or membrane. During further growth the larva loses the upper lip and the finfold, and the eyes and other organs grow into the adult lamprey.

(b) Gnathostomata

(1) *Fishes (Pisces):* The Pisces or Fishes proper, the next group to be considered here, will be dealt with in an independent chapter in view of the great importance of the group for mankind, among the inhabitants of the sea.

(2) *Higher Vertebrates :* Above the level of the fishes in the scale of evolution, only a few are really marine inhabitants. Among those are to be mentioned the sea snakes and turtles among the reptilia, and the whales, seals and porpoises as well as dolphins among the Mammalia as the most important members.

There are several species of turtles in the sea and these are useful for the food of man; some can grow to a length of 10-12 or even more feet at times. They are characterised by the possession of a strong shell of several bony plates fused together; the entire animal can withdraw at will into the shell for protection at the slightest sign of danger from outside. Even in normal position the entire trunk region is enclosed inside the shell and only the limbs, head and tail protrude outside of the shell. Some turtles can travel long distances in the sea to reach the shore for egg-laying. Just at the required time the creatures are guided by mere instinct to reach some shore at any cost and then dig pits in sand to deposit the eggs there. The turtles and sea snakes are both air breathers and possess lungs like the land animals themselves and they must invariably reach surface of the water at longer or shorter intervals of time for taking in gulps of fresh air. At least five species of turtles seem to be common in the Indian seas.

Many species of birds such as the gulls, terns and albatross live over the surface of water in the sea and hundreds of others fly over the sea during certain seasons of the year. There are species which have lost all powers of flight and there are some which have only limited flying powers; and there are birds like the penguins which are entirely dependent on the sea for their habitat. In spite of all this, birds are essentially aerial animals and their importance in the overall biology of the sea is somewhat limited in spite of the economic importance of some guano-producing birds of some coasts (e.g. Peru).

The whales are among the largest of living animals and are also a source of profitable fishery for many countries which share this resource internationally. The whales are mainly animals of the oceanic regions and of the temperate and semi-temperate climates though some whales move about quite widely and dead whales are occasionally washed ashore at various parts of the Indian coast. India has no whaling programme at present but it is quite possible that in some near future we may also own some whaling ships and take our share of whales from the international whaling areas in different parts of the sea. There are different species of whale and they seem to live mostly in communities so that for fishing them the fleetmen concentrate over a particular known area of the sea and carry out their operations. Fishing of whales is done usually by means of harpoon guns and spotting out the areas of their occurrence is done by looking for certain characteristic fountains of water that they keep blowing out now and then from their blow holes or nasal apertures.

Most whales feed on plankton and one category known as the baleen whales possess elaborate filtering plates in the mouth for sieving the water and collecting the organisms for food; these plates are called the baleen plates or "whale bones" and are large fanlike expansions of intertwined fibrous material.

Seals and dolphins are also subject to some fishery here and there in some seas, the dolphins being quite common around our coasts. However, there is neither a regular fishery for dolphins in our seas at present nor do we yet possess any extensive and reliable knowledge about their behaviour and life-histories. The common dolphin of our waters belongs to the species *Delphinus delphis* and has jaws which resemble birds' beaks. This form has often been reported preying on shoals of small fish such as mackerel and the oil sardine.

Another mammal deserving mention here is the seacow which belongs to the order Sirenia while the whales belong to the Order Cetacea. The common seacow is *Dugong dugon* which seems to be not uncommon in some parts of our seas such as the Gulf of Mannar; a pair of these dugongs were kept alive for ten years and longer in an aquarium tank in the Central Marine Fisheries Research Institute Laboratories at Mandapam Camp, this being perhaps the longest period of their known captive life. They are "vegetarians".

Lalmohan, Nair and Rao (1975) have done valuable work on this dugong and James has given detailed list of particulars of whales all found stranded along the Indian coasts from 1748 to 1980.

Chapter IX

Fishes and Fisheries of Our Seas

A. THE FISHES

(1) General Introduction

Fishes are cold-blooded vertebrate animals with well-formed jaws; they possess an elongated chain of bones called the vertebrae and this forms the backbone or vertebral column of these animals. However, in the lower fishes of the group Elasmobranchii the skeleton is made up of a softer cartilaginous material and the true bones make their appearance only in the true fishes of the Teleostomi (which are therefore called the bony fishes in contrast to the cartilaginous fishes). The entire form and structure of a fish seems to be an adaptation to its aquatic existence, the spindle-like shape of the body itself being the most splendid adaptation for easy movement in the water in which the animal spends all its life. The body of the fish is also covered with a slimy secretion from the skin and this helps in the smooth movement in the water by reducing resistance to the medium surrounding it. The skin is protected by scales which are plate-like structures of horny or chitinous material arranged in imbricating or overlapping rows, the hind ends of the scales of each transverse row covering the front ends of the scales of the next row behind, and the anterior ends of the scales being themselves hidden in the scale-pockets inside the skin. The slimy layer of epidermis covers these scales all over.

The existence of special structures called fins is another special feature of the fishes, these being the organs of locomotion in the group, being both structurally and functionally well-adapted for this purpose. The fins of fishes are of two categories: (a) the median fins which are unpaired and situated along the median line of the body dorsally and ventrally; and (b) the paired pectoral and pelvic or ventral fins situated on the sides of the body at different regions, the pectoral generally corresponding in position to what may be called the shoulder region, while the pelvics or ventrals are in the pelvic region of the body. The median fins are developed from a single median fin-fold of the early embryo running as a continuous structure dorsally, then around the tail, and then again for some distance ventrally. The number of dorsal fins and the entire fin formula including the numbers of rays (i.e. skeletal elements inside the fins) in the different fins are of taxonomic value; some groups or families have a single dorsal fin while others have two; and there are a few like the Gadidae (cod fish and relatives) which have three dorsal fins also. The fins may have various modifications; in some fishes, one or more of the fin rays may be modified into spines; in a group of fishes called the gobies (Gobiidae) the two members of the pair of pelvic or ventral fins are fused together to form a cuplike structure for adhesive purposes, that is to stick firmly to substratum when strong water currents threaten to wash them downstream in the rivers where they live. In the Polynemids the pectoral fins have some of their rays highly prolonged. It is of course yet not possible to state always the exact functions of some of the peculiarities of structure noticed here and there in the fishes. Other peculiarities include the following: one or more fins may be absent; the caudal fin may have various shapes and structures; and parts of the fins may be divided into independent finlets; the dorsal fin may get modified as a sucker like structure; and so on.

The *Cyclostomata* are not true fishes though they are also treated in the Class Pisces or Fishes, being fish-like animals of the jawless *Agnatha*, that can be considered as the forerunners of the fishes. The true fishes *sensu strictu* consist of the Elasmobranchii and the Teleostomi. The former are the cartilaginous fishes including the sharks, skates and the rays, while the latter are the true fishes or bony fishes which form the vast majority of the fishes living at present in both seas and freshwaters. In the bony fishes also the skeleton appears at first as cartilage which is permanent skeletal substance of the Elasmobranchs, but this cartilage later gets ossified or converted into bone and overlapped by it in later life; the bone is a hard structure without any flexibility except where it is jointed as in some cases. The Elasmobranchs are sea-dwellers. These include the sharks, skates and the rays, all of which have only a cartilaginous skeletal structure.

Apart from the above features the most important adaptation of fishes to their aquatic existence is the presence of special structures called the gills - feather-like outgrowths—which are lodged in pouches or pockets in the pharyngeal wall, with slits opening to the outside; these slits are separate from one another and single for each pouch in the Elasmobranchs but they are united into a single large slit, the opercular or gill slit opening out from a common opercular chamber in the higher fishes. The gills are highly vascularised and as the water current passes over them during its passage from the mouth to the outside through the gill slits the oxygen of the water is exchanged for the carbondioxide of the blood in the capillaries of the gills and thus respiration is effected. This exchange of the gases occurs by diffusion through the thin permiabe membrane that separates the blood in the capillaries from the water outside. In *Hexanchus* and *Heptanchus* there are seven and eight pairs of gill slits respectively.

The covering over the gill slits in the bony fishes is the operculum and the chamber covered by the operculum opens out behind the head of the fish by means of the opercular slit. The gill slits open out directly and individually as already mentioned, in the rays and skates and also in the sharks, all of which are members of the Elasmobranchii.

The rays (Rajiformes) have a flattened body with the fins of the shoulder fused with the disc-like body; these are bottom fishes living in the shallow waters as well as in the deep seas. The skates have a body form intermediate between the flat-bodied rays and the sharks which are provided with a typically fusiform and slightly laterally compressed body. The sharks include some of the largest living animals except the whales. The largest of the bony fishes normally reach only a few metres in length though there are a few forms in some specialised groups reaching still larger sizes (such as *Istiophorus gladius* and *Makaira makaira*), which normally occur in the oceanic regions only.

Another speciality of the fishes is the occurrence of some sensory organs situated in the form of a canal system called the *lateral line canal system*; the sensory organs forming this system help the fishes in perceiving the changes in the pressure of the medium and certain happenings in the surroundings, all these perceptions generally guiding the fish in its movements inside the water. The lateral line canals may be simple linearly arranged sense organs, one linear set on either side of the body, or the canals may be variously branched and anastomosed with one another as in some of the abnormal flatfishes (Seshappa 1970).

When the sea is compared with freshwater, the following may be mentioned as the outstanding differences between the two types of environment: (1) The freshwater environment is restricted in area as well as in depth; whatever be the magnitude of the lake or water-area usually it can never be comparable in the magnitude even to a limited part of the sea in general; (2) the seawater has a higher density than the freshwater and contains, various salts and chemicals (that are normally absent or present only in minute quantities in some cases, in the freshwater) in solution as well as in suspension; (3) the waves, tides and internal currents occurring in the sea are almost absent in freshwater.

The fluidity of the medium and the availability of food and other requirements for living for the inhabitant organisms are the main common features between the two watery media, namely the sea and freshwater; fishes and other animals of same or similar groups are found in both environments. In both the media the animals are functionally and structurally adapted to the aquatic existence. Between the inhabitants of the estuaries, the shallow seas and the freshwaters there is often not much structural difference though the taxonomic differences according to the families to which they belong will naturally be prominent. Structural and physiological adaptations are however, inevitable in particular environments such as in the hill streams of freshwaters (where the gobioid type of the cup-like development of the pelvic fins is seen) and deep sea areas, or where a permanently evolved bottom existence has come into being as with the flatfishes with a somewhat twisted body-arrangement and with both eyes on the same side of the head in a distinct asymmetrical arrangement. In the freshwater forms such as *Clarias, Saccobranchus, Anabas* and *Macropodus* secondary air breathing organs are developed as they permanently inhabit oxygen-poor environments and the normal gills fail structurally as well as functionally to reach the required level of efficiency.

(2) Classification

The Class Pisces can be divided into three super-Orders or Sub-Classes as follows: (i) the Cyclostomata, (ii) Elasmobranchii (Chondrichthyes), and (iii) Teleostomi (Osteichthyes). We may now deal with these three groups briefly.

(i) The Cyclostomata: These are the jawless craniata distinct from the true fishes. The hagfishes and lampreys already mentioned constitute this group. They are defined as the "round-mouthed fishes" with the following distinctive characters:- (a) A long rope-like notochord is present along the dorsal aspect of the body above the alimentary canal; (b) the mouth is circular and has no jaws, the name Cyclostomata being derived from this character; (c) the skeleton has no bones but is made up of only the softer material called the cartilage (d) the gills have no supporting elements but there is a cartilaginous network called the branchial basket outside the gills providing a connection with the skull; (e) there are no paired fins; and (f) there is only a single median nostril on the head instead of the paired nostrils of true fishes.

(ii) The Elasmobranchii (Chondrichthyes): The Chondrichthyes are sometimes recognised as a larger group than the Elasmobranchii so as to include the latter sub-class as well as the Holocephali as a separate sub-class of the same bigger group. The term Chondrichthyes means cartilaginous fishes. The Super-Order Chondrichthyes has the following as the principal distinctive features of its constituent animals:- (a) there is a notochord which is like a string of beads, (b) the mouth is provided with regular jaws, (c) the vertebrae and the entire skeleton even in the adult animal is made of cartilage only; (d) the gills have a cartilaginous internal supporting structure (e) the skeletal supports of the gills have no connection with the skull in any way; (f) paired fins are always present; (g) nostrils occur as paired apertures found one on either side anteriorly on the head unlike the Cyclostomes where there is only a single nostril situated medially on the head.

The Elasmobranchii are the sharks, skates and rays and have the following distinctive common characters:- (a) there are five to seven pairs of gills and gill-slits; (b) there is an aperture called the spiracle in front of the row of gill-slits; (c) there are no scales comparable to the scales of other fishes but there are small placoid scales or dermal denticles imbedded in the skin, these themselves becoming enlarged as the teeth on the jaws; (d) the genital opening and the anus open into a common cloacal chamber which has a single aperture to the exterior; the undigested remains from the alimentary system as well as the genital products pass out through the common opening of the cloaca; (e) the males possess a pair of copulatory organs the *claspers*, situated near the cloaca.

The Holocephali are also known as the chimaeras and have the following distinctive features: (a) there are four pairs of gills but a single pair of gill-slits; (b) there are no scales in the skin; the anus and the genital opening open out to the exterior separately, there being no common cloacal chamber as in the other Elasmobrachii; (c) a pair of claspers are present in the male to assist in copulation, and often there is also a frontal clasper as in *Chimaera monstrosa*.

(iii) The Osteichthyes: These are the most numerous of the living fishes of the present time; their chief distinguishing characters as compared to the Chondrichthyes are:- (a) the vertebral column is made up of distinct bony vertebrae; (b) there are two jaws, the upper and the lower guarding the mouth as in all higher vertebrates; (c) all skeletal elements are made of true bone (except in the embryo) and the gill-arches are also bony; (d) the gill arches have no connection with the skull; (e) paired fins are present in addition to the median ones; (f) The nostrils are also paired apertures.

Two main sub-classes are recognised under the Osteichthyes namely, the Sarcopterygii and the Actinopterygii. In the Sarcopterygii the fins are thick and muscular and covered with scales and also supported by strong finrays; the nasal openings may be external as in the lung fishes or internal as in the lobefins. In the Actinopterygii the paired fins remain as in the above and the nasal openings are external.

(3) Some details on true fishes

Of the various main groups mentioned above, a few remarks may be now given about the Chondrichthyes and the Osteichthyes, leaving out the hagfishes and lampreys which have already been referred to very briefly earlier. Of the bony fishes the group Dipnoi or lung fishes do not seem to occur in Indian waters.

(a) The Chondrichthyes (continued): The Elasmobranchii and the Holocephali are the two important sub-classes here, the Holocephali being however, a small group with many extinct forms and usually treated as part of the bigger sub-class Elasmobranchii. These fishes seem to have been very well-represented during the past geological ages. At present the important genera are *Chimaera, Callorhynchus,* and *Harriotta. Chimaera* occurs in the seas of Europe, Japan, Australia and North America and also in the seas off the Cape of Good Hope of South Africa. *Callorhynchus* is somewhat plentiful in the southern temperate seas while *Harriota* is a deep sea fish. Very recently a fish of this group named *Neoharriota pinnata* has been described from our seas by Silas (1969).

The Elasmobranchii is a fairly large group with several living species of rays, skates as well as the sharks. After the whales the sharks are the largest of the marine animals at the present day. The tiger shark or blue shark *Carcharhinus* and the whale shark *Rhyncodon typus* are among the largest of sharks and occur in the Indian seas also, more frequently on our east coast where the sea deepens relatively at a shorter distance from the shore than on the west coast. Many of the large sharks are denizens of the open oceans while others may occur at all depths including the shallow inshore waters also. The main morphological difference between the sharks and the rays is in the form of the body, the rays being also mainly bottom-dwellers while the sharks move about at all depths. The body of a shark is divisible into the head, trunk and tail regions and has the typical fusiform or spindle-like shape. The body is slightly laterally compressed as in the generality of fishes. A common inshore example of sharks is the dogfish *Scoliodon sorrakowah* (Fig. 49) which is usually given for class work in most colleges in India. The larger sharks have also a similar shape of the body, the main difference relating to the fin positions, colour and the proportions of the different parts of the body to one another. The fins of sharks consist of two dorsals, placed one behind the other, paired pectorals, and pelvics (or ventrals) and a single caudal (which is median and heterocercal with a longer upper and a shorter lower lobes), and an anal fin. The caudal fin is laterally compressed and all the fins are thick and adipose. Variations from this general formula of number and structure exist in some species. The hammer head shark has the head shaped like a hammer with the two eyes situated at the two ends of the hammer-like head.

While the colour of the sharks is variable the most frequent shade is a basic brown or greyish brown. Stripes and spots are frequent in many species. Sharks live by feeding on other animals of the sea, especially the fishes. Shark fins are used for making "shark fin soup" which is a delicacy in some countries, and the liver of sharks yields medicinal oil fit for cattle as well as for human consumption. Shark flesh is edible and fetches some value though it is not always specially relished. Shark skin is useful in some industrial applications.

The sexes are separate and fertilisation is internal in the sharks; the fertilised eggs are surrounded by a case of horny material secreted by the uterine wall and all development goes on inside the case; viviparous reproduction and ovo-viviparity are also known in some species; the egg-case of sharks is bag-shaped and is called the mermaid's purse. In ovo-viviparous cases eggs are retained in the egg-cases for a longer period.

The rays and skates are also close relatives to the sharks but they have the head and trunk fused together and only the tail is distinct from the rest of the body in external appearance. The pectoral fins are fused with the trunk so as to provide a broadened shape to this part of the body. Some species have spines on the tail. One kind of rays called the electric rays have a group of cells in the body that can generate electricity of a low voltage and provide mild shocks to any person touching them when they are alive.

Among the skates there are many trawl fishes of good market value in the Indian seas. *Rhynchobatus djeddensis* is a good example of these and grows to several metres in length. *Pristis* (Fig. 50) is a genus of skates and is popularly known as the saw-fish with at least three species in the genus, the common one being *P. cuspidatus*. In these species the upper jaw is prolonged and developed into a saw with a row of strong sharp teeth on either side, the entire structure serving as an appendage for offence and defence. The external differences between the three species of the saw fishes relate mainly to the size, number and arrangement of the teeth on the jaw, which is developed into the saw. The skates do not have a body so wide as in the rays in relation to length, as the fins here are distinct but the trunk is flattened unlike the sharks, the posterior region of the body looking however like the sharks only. Thus the skates may be said to have an intermediate type of body appearance between the sharks and the rays. The sawfishes belong to the family Pristidae and the skates belong to the family Rhinobatidae. One of the skates *Rhina ancylostoma* has a head rounded anteriorly and the general shape of the body resembling roughly the musical instrument, guitar; hence this fish is called the "guitar fish".

Dasyatis sephen is a common species of rays in our waters; *Dasyatis bleekeri* grows to a length of 8-10 feet but most of this length is contributed by the long tail; the rays otherwise seem to grow to a few feet only in length, though some giant rays are known of relatively large sizes. *Mobulus diabolus* has a wider body diameter (upto 18 feet) than the length in the fully grown fish. *Aetobatis, Rhinoptera* and *Aetomilus* are some other examples of rays from our waters.

(b) The Osteichthyes (continued): Of the two groups mentioned under the Osteichthyes the Sarcopterygii are only a small group; these possess thick fins covered with scales and include the Coelacanths exemplified by *Latimeria chalumni* discovered and worked out by the famous ichthyologist, the late Dr. J.L.B. Smith of Rhodes University in South Africa. Until this discovery some decades ago, it was generally believed that the Coelacanths were completely extinct. These fishes form a link between the modern fishes and certain extinct groups of fishes.

In the Actinopterygii the fins are not thick and limblike as in the above fishes but are thin and supported by horny finrays. These are *sensu strictu* the modern bony fishes.

According to Misra (1959) there are about 402 species of fishes which are of commercial importance in our country taking both marine and freshwater forms into account and leaving out the less economic forms. However, if all forms are included we may have well over two thousand species in our country, the marine species being perhaps more numerous than the freshwater ones in this number. Without going into details of the different groups we may mention the following as the more important sub-Orders of marine fishes of the Teleostei: Symbranchiformes (two families); Anguilliformes (eels); Pleuronectiformes (flat-fishes); Echeiniformes (Suckerfishes); Polynemiformes (threadfins); Thunniformes (Tunas); Tetradontiformes (Puffers and Globefishes); Beloniformes (including half-beaks and others); Bericiformes (including the Holocentridae); Mugiliformes (Mullets); Perciformes (Perches and related fishes); Scopeliformes (including the Bombay duck *Harpodon nehereus*); Clupeiformes (including the famous Hilsa and such other forms) and Batrachoidiformes (the frog fishes).

Various characters of the body are taken into account in the classification of the fishes. The following may be mentioned among the main features considered: (a) form and size of the body, (b) number, shape, size and other characters of the fins including also the numbers of finrays and the finspines when the latter are also present, (c) the distance of the different parts of the body from the anterior end, and the dimensions and proportions of the different regions of the body, (d) the size, shape and structure of the scales covering the skin of the body, (e) the numbers of scales along the lateral lines and across the body at specified areas, (f) the size and positions of the eyes (g) features and positions of the nostrils, (h) size and peculiarities of the jaws and the number, size and arrangement of the teeth on them, (i) the numbers and size of the gill-rakers, and peculiarities of the operculum (gill cover), and the pre-operculum (plate in front of the gill cover), (j) presence or absence of air-bladder, (k) numbers of pyloric caccae, and (l) various other internal and external features.

It is because there is a great deal of variability in all these characters individually and also in their various combinations that there are so many families among the fishes, and that classification is not by any means always an easy matter. There are many cases where long personal experience alone can guide in differentiating one species from an allied one and there are occasional instances where even long experience is not entirely helpful without further studies in the light of any latest knowledge gained in the field everywhere in the world.

Though only the external characters are used for identification of the various forms in general practice in fisheries work, there will naturally be correlated internal differences and resemblances also among the various groups at all levels, and any natural classification must result in the arrangement of the different families, genera and species in the order of their natural relationships from the evolutionary point of view. The determination of the various genetic and evolutionary relationships among the different groups is done after a detailed consideration of all aspects of representative numbers of species in the different families. Revision of arrangements already made after further studies is not also ruled out and in fact there are frequent instances of differences of opinion among workers in systematics regarding the taxonomic or systematic positions of certain species and even higher groups.

Many variations of body form in fishes may be related directly to the present day mode of life of these animals or a particular character may be a retention of the same developed during evolution from a past form or structure. The eels (Anguilliformes) have an elongated and snake-like body; some of these can swim rapidly and cover long distances. The Atlantic eel has a breeding migration covering thousands of miles in its life history and this aspect has been the subject matter of a life time research by Prof. Johannes Schmidt (1922 etc.). The Pleuronectiformes are very peculiar fishes in regard to body form; these are the flatfishes; it will be remembered that the body is flattened in the rays also among the elasmobranchs, but there, the bilateral symmetry of the body is not at all lost or altered; but in the flatfishes of the Pleuronectiformes (families Psettodidae, Bothide, Cynoglossidae etc.) or Heterosomata as they are also called, the bilateral symmetry is lost in the adults in which there is such marked asymmetry that even a rough and distant glance at these animals will impress anybody that these are indeed peculiar creatures. These flatfishes normally live at the seabottom and it may be said that the shape of the body is an adaptation to this bottom life of these animals. In larval life these fishes have a symmetrical body like the other fishes but during metamorphosis one of the eyes migrates to the other side, the head and body undergo a certain amount of twisting and other changes, all these resulting in both the eyes being finally on one side of the head (Fig. 43); the fish lies on the seabottom on one side of its body which has the longitudinal fins (dorsal and anal) along its margins; the mouth, jaws and nostrils also show out the asymmetry. The eyes are on the upper coloured side of the flattened body while the lower side which is white is blind. The various shades and patterns of colour are found only on the upper eyed side of the body.

There are at least five families and 92 species in the Heterosomata in our seas (and freshwaters) as listed by Norman (Norman 1928). *Cynoglossus semifasciatus* (Fig. 51) is a fish of considerable commercial importance on the west coast, particularly off coasts of North Kerala and South Karnataka, forming a good seasonal fishery in this area. This and other flatfishes are important not only because of their fishery value but also from the scientific point of view because of the various peculiarities of structure, life history and adaptations exhibited by these creatures in the sea.

The Echeinformes includes *Echeinis* and *Remora* which are known as the suckerfishes; these frequently move about attached to the undersurfaces of the bodies of other bigger fishes; for this attachment the dorsal part of the head region is provided with a strong oval muscular sucker-like structure which is very powerful and efficient for this function.

In the Polynemiformes the most important are the fishes of the family Polynemidae; *Polydactylus indicus* and *Polynemus tetradactylus* are the famous "dara" and "rawas" of the trawler landings at Bombay; these grow to very large sizes (among the bony fishes) and fetch good prices in the market, particularly the "rawas" (or "Indian Salmon") which is considered very tasty by the fish-eaters. The most outstanding structural feature of these fishes called also threadfins, is the presence of thread-like prolongations of some of the pectoral finrays, their exact numbers varying from species to species; *P. tetradactylus* has four such elongated rays while *P. indicus* has five and *P. heptadactylus* has seven.

The tunas (Fig. 53) of the family Thunnidae of the Thunniformes are also economically very important fishes of the sea. These fishes are caught in abundance only in a few centres along our coasts. Among these centres may be mentioned the Karnataka coast north of Mangalore, the Laccadive sea around the Lakshadweep Islands and Tuticorin along the east coast; there is also a good tuna fishery in the sea around the Andamans. The tuna fishery has been a thriving industry in the Lakshadweep Islands for a very long time. The islanders preserve the tunas in a dry form which is called "masmeen" in the market and has been known to keep for indefinite lengths of time. In recent classifications the tunas are included along with the mackerels in the order Scombriformes.

Cybium is the seer fish with four Indian species all being of high table value, much relished by eaters and highly priced. *Rastrelliger kanagurta* (Figs. 41a & b) is the Indian mackerel belonging to the family Scombridae. This fish and the oil sardine together contribute about 40 to 60 per cent of the sea fish production in India every year. One of the questions engaging the attention of the mackerel specialists for some time has been whether there are different races in this species or not. The reason for this curiosity is the fact that between the east coast and the west coast there are considerable differences in the seasons of occurrence, growth trends, spawning times and other features in the life history of the fish. Moreover variations in fishery and other features exist in the species among different centres of the same coast also. Dhulkhed and Rao (1976), Rao and Dhulkhed (1976) and Dhulkhed and Nagesh (1976) have proved by means of biochemical and serological studies that there are different races in the mackerel (*Rastrelliger kanagurta*) occurring at Mangalore and in the immediate neighbourhood. Seshappa's studies (1983) have indicated (from the frequency of scale-circuli in different seasons and years) that there is no change in the stock of this fish at Calicut over the period of several years from 1969 to 1976.

Among the Tetrodontiformes there are frequently poisonous forms, the spines on the body being the sources of the poison. In several species the spines of the fins may be poisonous. *Tetrodon* has a puffed-up globular-looking body which may be actually due to its being filled with a considerable quantity of air; *Balistes* is a fish harmful to smaller fishes of the fishing grounds as it preys on them.

The Perciformes is a very large order with as many as 45 or more families, the true perches being members of the family Percidae of this Order. The Sub-Orders of the Perciformes are: the Trichiuroidei, the Gobiodei, the Stromateoidei, the Scombroidei and the Percoidei. The following families are included under the Percoidei: Cichlidae, Menidae, Plotosidae, Drepanidae, Carangidae, Lactariidae, Sciaenidae, Latidae, Lutjanidae, Leiognathidae, Lethrinidae, Nemipteridae, Ambassidae, Serranidae, and Theraponidae.

In the family Cichlidae the fish *Tilapia mossambica* has been useful in fish cultural practices and has given quick results, these stimulating work on other and more useful species. the Pearl spot, *Etroplus suratensis* also belongs to this group. The Menidae are small fishes; a member of this family, *Mene maculata* has been occurring in large quantities in the trawler catches off Calcutta. The batfish *Platax vespertilio* belongs to the Platasidae. This fish has a coloured and banded body with a highly laterally compressed but vertically exaggerated shape. *Drepane punctata* is a common fish of the Drepanidae and is known as the moonfish and Chand (in Hindi), perhaps because of the rounded profile of the body which is laterally compressed and rather flattened.

There are several food species in the family Carangidae, *Megalaspis cordyla*, a fish which is looking somewhat like the mackerel, belongs to this family. There are various species of *Caranx* contributing to fisheries of small magnitude along different parts of the coast, *Caranx crumenophthalmus* being the horse mackerel. The Lactariidae have the species *Lactarias lactarias* as the representative. The Sciaenidae are an important group with over twenty species in India, with some of these growing to as great a length as 5-6 feet and reaching a weight of upto 150 lb. or more. There are some species of this family forming a common fishery along the west coast. Of these species *Pseudosciaena diacanthus* (=*Protonibea diacanthes*) (Fig. 48) and *Otolithoides brunneus* (=*Otolithoides biauritus*) are economically the most important in the sea, fetching good prices in Bombay as well as elsewhere. *Otolithes argenteus*, the silver bar fish is a common member of the same family occurring in most parts of the coast. There are also some freshwater species in this family such as *Pama pama*. Lal Mohan (1969, 1972 and 1981) has made a thorough revision of this family using the airbladder among

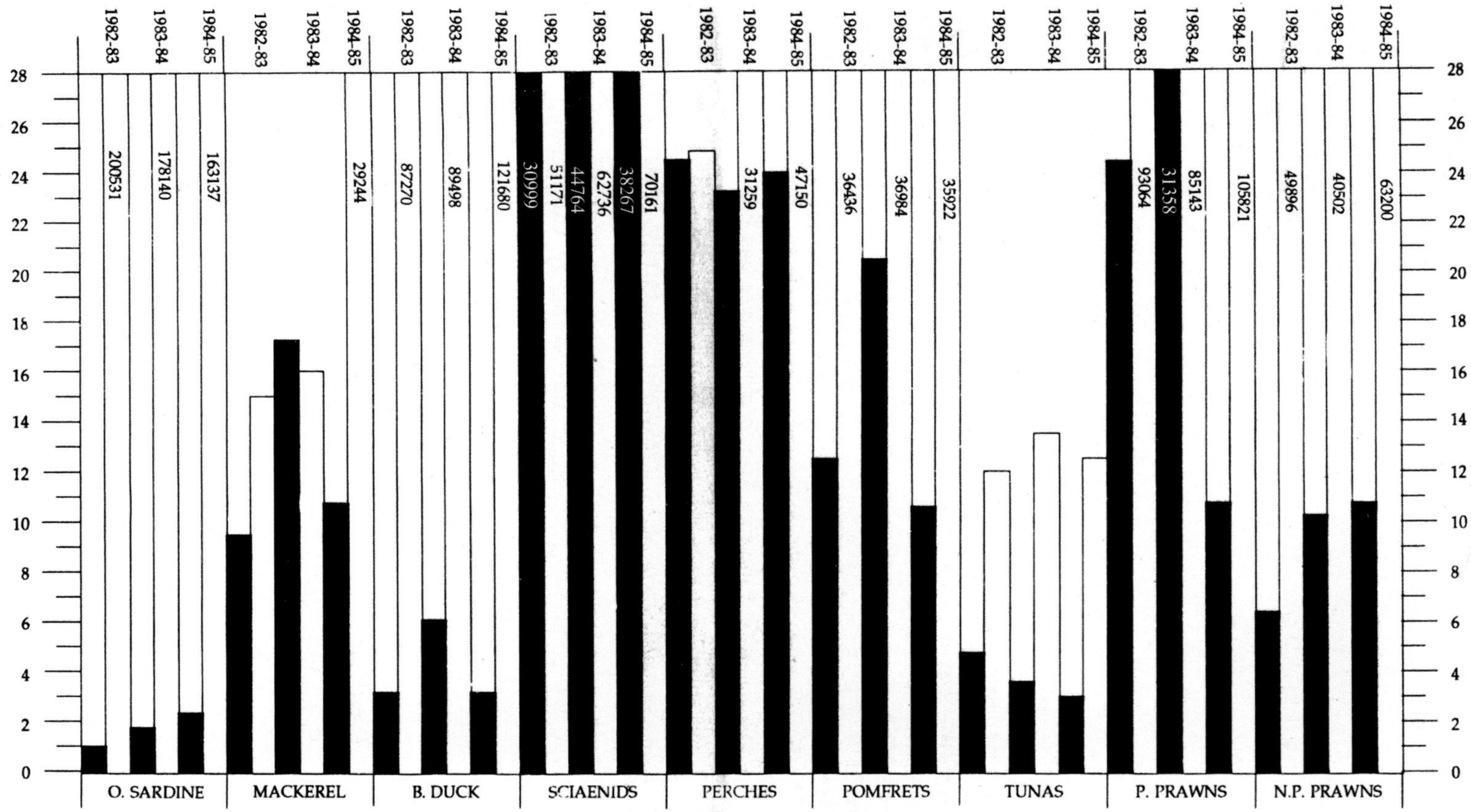

Fig. 38. Estimated Annual Total Landings of nine categories of sea fishes from the east and west coasts of Inida during 1982-83 to 1984-85
(Source of data: *Mar. fish infor. Serv. T. & E.. Ser.* No. 67, 1986, CMFRI, Cochin.)

E. COAST
W. COAST

LANDINGS IN THOUSANDS OF METRIC TONS

the important characters used in the classification following Trewawas (1962 and 1965) and has recognised as many 36 species from our waters.

In the Latidae *Lates calcarifer* grows to the large size of upto five feet. In the Lutjanidae there are several marine species *Lutjanus johnius* and *L. roseus* generally occur in good quantities along with the sciaenids in the sea off Kutch and elsewhere. The Leiognathids are small fishes occurring in variable quantities in the sea from place to place, some of them occurring in the freshwaters also. The Lethrinids are known as the breams one kind of these being called the "pig-faced breams" because of the peculiar appearance of the head reminding one of the front part of a pig's head. The serranidae also include fishes that can grow to a large size; these look very much like the sciaenids in general appearance but with important differences.

The catfishes (siluridae) are another important group among the seafishes there being again some "catfishes" in the freshwaters also. Some common species of catfishes yield good catches at some centres. Their skin is without scales and their lips possess tentacular sensory processes called the barbels.

The popular pomfret belongs to the family Stromateidae and occurs fairly abundantly at Bombay, Ratnagiri and other areas. Three common species are recognised, namely, *Pampus argenteus* the silver pomfret, *P. cinereus* the white pomfret, and *Parastromateus niger* the black pomfret which is also equally abundant and quite popular as food in some places as along the Kerala coast. The silver and white pomfrets are often air-lifted to Delhi and other inland centres (when abundant at the coastal centres) for marketing there, in view of their popularity as table fishes and the good price that they fetch in the interior urban regions of the country.

The swordfish *Xiphias gladius* which grows upto 15 feet in length belongs to the Xiphiidae while the daggerfish and the sailfish (*Makaira* and *Istiophorus* respectively) belong to the family Istiophoridae. All these fishes of the various families mentioned in the last part of the above section belong to the order Perciformes as already mentioned.

B. MARINE FISHERIES

(1) General:

It has already been pointed out that the fishes are the most important among the animals of the sea from the point of view of their utility to mankind; though fisheries activity has progressed a great deal in recent years, the profession of catching and selling fish has always been there from times immemorial. As the human beings of the past were mostly non-vegetarians and fishing in shallow waters of the sea near the shore was the easiest thing to do for food, men from ancient times were perhaps taking recourse to this method as a natural course open to them in their daily food-finding attempts. It is easy to imagine that in the earliest times there might have been no general restrictions on fishing either in the sea or in the freshwaters, except when the latter were part of

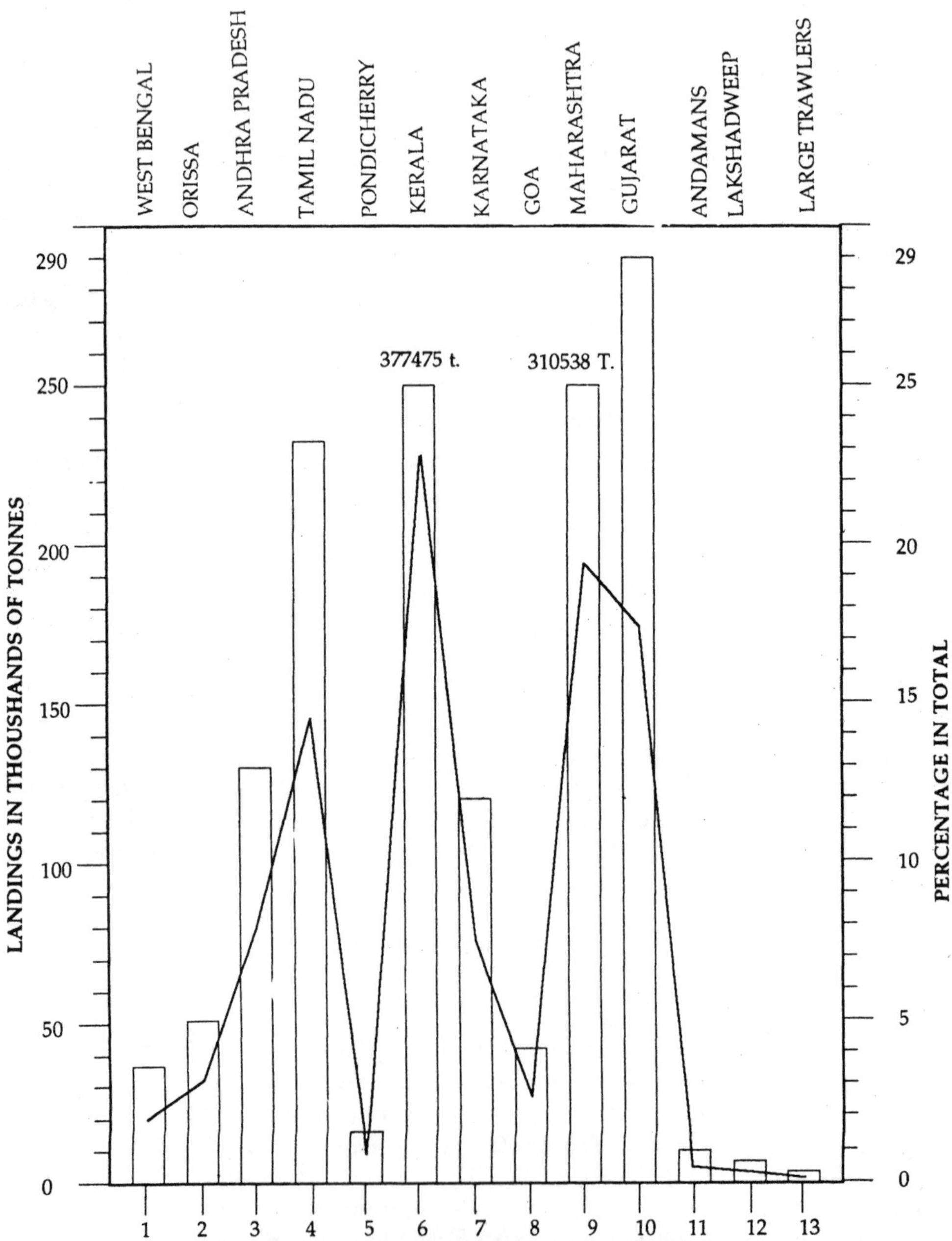

Fig. 39. Total Landings of Seafish in Different Maritime States with Percentages (darkline) in the All India Totals during 1984 (*Source of data: Mar. Fish. Infor. Serv. T. & E. Ser. No. 67, 1986, CMFRI, Cochin*).

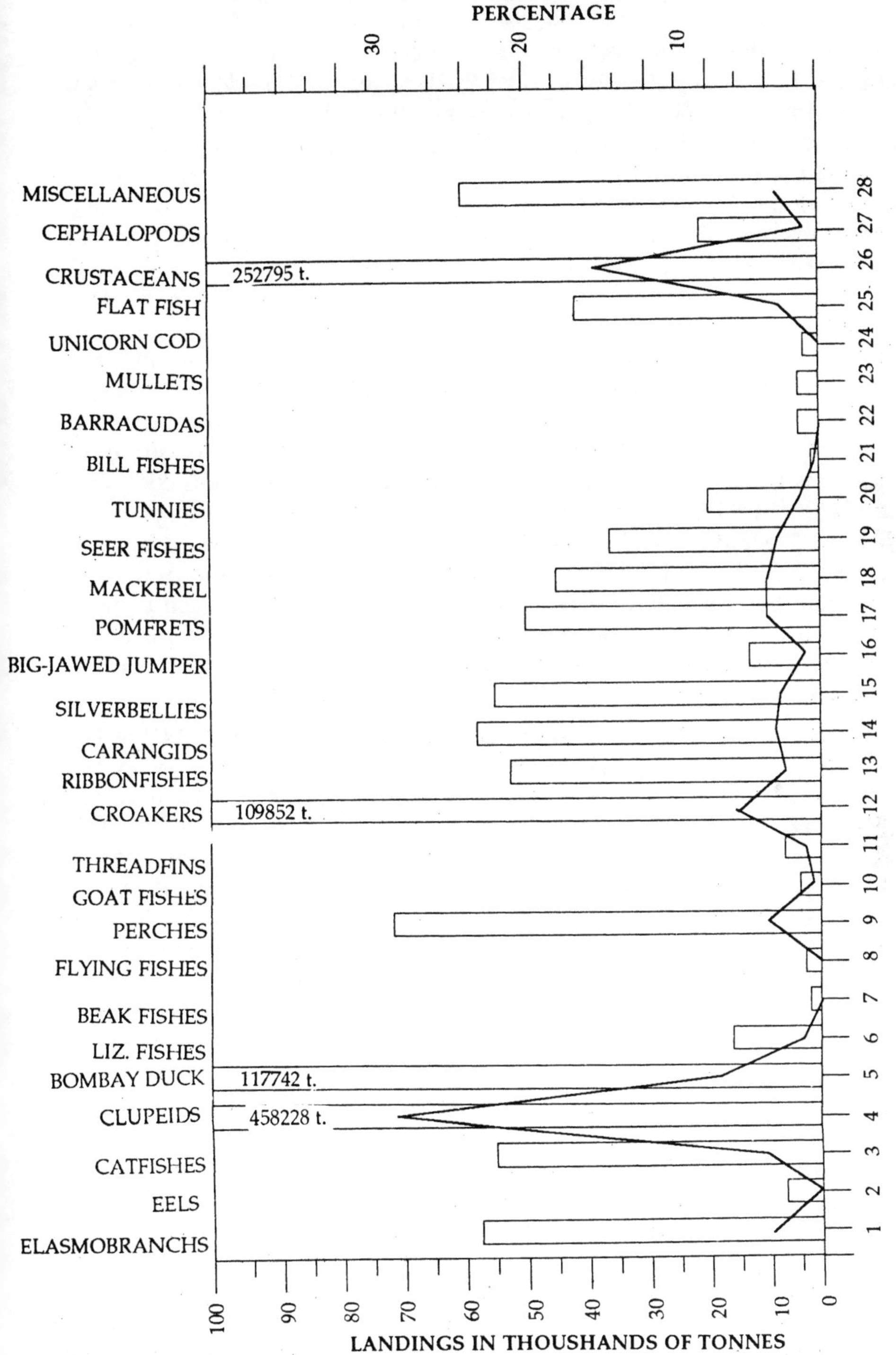

Fig. 40. Estimated All India Fish Landings from the sea during 1984 with percentage of each category in the total landings. (Source of data: *Mar. Fish. Infor. Serv. T. & E. Ser., No. 67, 1986*).

personally owned properties. With the passage of time, social and governmental control might have been found necessary and enforced in some form or the other here and there. So far as India is concerned there seems to be evidence that even in Emperor Asoka's times there were fishery regulations enforced by the Government within the areas of its jurisdiction. These regulations are said to have included the prevention of fishing in certain areas of water and in certain periods such as the newmoon and fullmoon days. There can be no doubt that these restrictions would have been thought of as helpful as results of experience and expediency in the interests of the people concerned, so that their fish resources might last longer and might be used more economically. In modern times, with the progress of scientific researches and the added knowledge about the fish as well as the fisheries, regulations have been introduced in most countries of the world according to indications and requirements from time to time. In the sea however, the introduction of such regulations was naturally rather slow in view of the vastness of the resources and the want of sufficient knowledge about them even in recent decades; experimental introduction and temporary lifting of rules have also been tried now and then in various places to see the results. The main thing in the present period is that scientific findings and principles are being applied more and more in the fishing practices with the aim of a better scientific utilisation of the fisheries all over the world, each country having control over a particular area of the sea along its coastline and the ocean outside this area being open to international exploitation by common consent among the nations of the world.

It may be stated that the attempts at a scientific exploitation of the sea fishery resources became a reality in India more after the attainment of Indian Independence. Some foundation was no doubt laid earlier in a regional sense in that the Bombay, Bengal and Madras Fishery Departments had conducted a few experiments in deep sea fishing and also located some traces of the present offshore fishing grounds which are now found to be good areas for deep sea fishing; the trends in the stocks in these areas are now being permanently monitored by the Governmental sea fishery Research Organisations and these have been steadily expanding their activities from year to year.

There are differences between freshwater fisheries and the sea fisheries in certain ways: the species involved are of course different and the magnitude of the catches are also different from region to region; the sea has a vaster resource for exploitation by man by fishing and subject to certain natural laws this resource is self-sustaining and self-renewing. Fishing in freshwaters is also mostly a relatively simple matter and many of the traditional methods are quite sufficient with improvements even under modern conditions, except in the very large lakes and tanks where some mechanisation can be attempted. Stocking and culturing are the special features of the freshwater fisheries while such practice is largely absent in the sea except for the backwaters and inshore areas of the sea. Large-scale mussel culture and prawn culture in different regions of the Indian coast are coming up well in recent years only, but these may open up further work on all-round marine fish culture in the future. At present however, by and large the marine fisheries in India, as elsewhere in the world, are essentially capture fisheries and regulation or rational utilisation of the stocks will continue to be dependent upon restriction on capture in required centres at indicated periods and also on particular sizes or life stages as revealed by continued research results.

Sea fisheries may broadly be considered under two divisions: (a) the deep sea fisheries and (b) the shallow sea fisheries or inshore fisheries; this distinction is based mostly on the regions of fishing and the types of equipment used, these latter being related to the depths and areas of fishing. The facilities required for fishing are naturally dependent upon the region and depth of fishing, elaborate preparations being naturally required for work in the open ocean and at great depths. Biologically it may not be correct to treat these two categories of fish stocks as distinct and separate because, some species are of a migratory nature and depending on the seasons, the same species can be found in the inshore as well as in the offshore catches, though not always simultaneously. For an understanding of the biology and life histories of the species involved both the fishes and their environments will have to be sampled and studied in detail. It may however, be stated that a good number of our inshore species, especially the smaller ones do not seem to occur in any large quantities in the offshore waters as a rule, though exceptional instances of such cases may be noticed here and there. Then the most important of our inshore fishes namely the mackerel and the oil sardine are migratory and their occurrence in large quantities in the inshore areas (i.e the presently known fishery areas of the respective fishes) is purely seasonal. An integrated study on such species from all regions of their occurrence and in all seasons of the year is most important in fishery research programmes.

"The Central Marine Fisheries Research Institute has been functioning as a nodal organisation for the collection, analysis and dissemination of data on marine fisheries and living resources of the Indian seas. The Institute developed a stratified multistage probability sampling design to collect information on exploited resources of our multi-species fisheries wherein multiple types of gears are used. Altogether 90 field staff stationed at 41 centres spread along the coast of India are exclusively engaged in the collection of catch and effort data. This work is carried out throughout the year, and covers about 1400 landing centres in different maritime states. The field staff are also regularly collecting length frequency data on commercially important species. About 800 species of fish and shellfishes and other resources of the Indian seas have been coded, preparatory to computerisation of these massive data ....".

A large number of "scientific and technical personnel working at different research centres are engaged in the collection, processing and analysis of data, on various biological aspects such as size, weight, sex and maturity stages of commercially important species, and also environmental data. Several case studies on marketing and socio-economic aspects in the marine fisheries sector are being undertaken by the Institute. More than 40 scientific and technical personnel are engaged in data processing and analysis. The results of the studies are regularly published in Research Journals and the Institute's publications" (cited from Silas 1982, *Mar. Fish Inform. Serv.T. &E. Ser.* 46). It may be added here that all this gets more and more strengthened and methodologically and otherwise constantly improved as and when found necessary. Many technical and other improvements have been already made since 1982.

2. Fishing gears and nets

The seafisheries that have been exploited the world over from ancient times are only the inshore fisheries; this is the practice continued even now in many parts of the world over from ancient times, there being various indigenous and traditional gears and methods according to the fishes being caught. Our own fishermen are quite clever and have evolved through the ages from experience, very ancient techniques and gears for fishing various species according to the depths and seasons of occurrence in the sea. In Kerala for instance different nets are in use for different species at the peak season of each fish species; but when no particular species is dominant and the catches are of a miscellaneous type, there are other nets that are found more useful in catching all or most of the available forms at the time, there being thus a general net as against a specialised one. The commonest marine or freshwater net is for instance the stringed castnet which is a circular net with weights at the periphery and a number of controlling strings radiating from the handle to the periphery, the handle being held by the operator while casting the net on water. Holding the handle in one hand and collecting the entire net in the other hand in a prescribed way, the operating fisherman throws the net so as to spread out on the water surface in such a way that it covers the sighted fish (or fishes, at random when no shoals are sighted). The periphery of the net now sinks as a circle which becomes smaller as the net sinks lower and lower. By a clever manipulation of the strings running from the handle to the periphery of the net, the operator can so adjust the entire operation that by the time the net sinks fully the metal "sinks" or weights of the margin would have all come together making the net a nearly enclosed bag ready to be taken out of the water with the contents. Of course some fish get entangled in the meshes also but a good part of the fishes enclosed come up as the harvest when the net is pulled up.

This is a very skilful way of fishing and only the very experienced are allowed to operate the castnet when any serious fishing is done. During the shoaling season of the Malabar sole *Cynoglossus semifasciatus* the castnet is used in addition to the boat-seine when large shoals are found and both make good catches. Much of the success of this type of fishing depends naturally on the manual dexterity of the operator. These nets can be of various sizes and cotton is the normal material used for making the twine used in such nets. These days nylon nets are in vogue in most places and governments subsidise the purchase of the nets and associated fishing materials which can be otherwise very expensive.

The gillnet is another type of net used both in the sea and in large areas of freshwater such as the lakes and tanks. This net is made of a number of pieces of fixed dimensions, the length of the net being adjustable by attaching or detaching pieces as may be necessary; the mesh of the net can also be of various sizes (usually from 1/4 to 1/2 inch to several inches diameter from knot to knot). The entire net is suspended longitudinally by means of floats and weights so as to form a fence of wall-like structure in the path of the fish; the fish trying to cross the net get caught in the meshes of the net while thus swimming their way to cross the net. The height of the net and the mesh size can be various according to requirements.

Another common type of net is the seine net, the shore-seine being the common one in many places for use near the shore. The seine net is essentially a bag type of net, the size of the bag as well as the wings and ropes used in operating the net being all variable according to circumstances. The shore seine is a special seine net and perhaps the largest shore seine in the whole world is the *Rampani* used largely in the mackerel fishery along the Karnataka coast. It is a net with very long wings and requiring scores of men "60 to 80" to operate the same. After surrounding the large mackerel shoals the wings which are carried in a smaller boat are brought around the shoal and pulled by two co-ordinated teams of men for a long time (1 to 3 hours) so that at the end the" bag" of the net with its huge catch of the fish makes its appearance at last. An average of about 8 to 10 lakhs of mackerel can be caught out in a single rampani haul during the peak season, and there are exceptional instances of record catches of upto 20 lakhs and more of fish (weighing in all from 150 to 180 tonnes) being made in a single haul of the rampani at Karwar (in 1958 for example). Fishes like the mackerel can be impounded also in the same area of water leaving the catch surrounded by the net so that the fish remain trapped beyond escape but at the same time they remain alive for some days even in the impounded condition because of the oxygen supply and water being available to them; this measure is taken recourse to when there is a glut of the fish landings with the resultant fall in prices to a very low level; big losses are prevented to some extent by a controlled release of the fish to the market at regular intervals depending on the demand and price.

A net similar to the rampani but smaller that it is the *Kairampani* also called the smaller shore seine (which it actually is). This and its modification are in use at various places on the west coast especially along the Karnataka coast. It is used very near the shore.

Another bagtype net which is also a seine net is the boat seine or *"Paithuvala"* of the Kerala coast. This is a much smaller version compared to the rampani described above, the wings being very short. This net is operated as a two-boat unit with six to eight men operating the same, three persons manning each of the two boats. After spreading the net the two boats drag on the wings (one held by each boat, and the two wings ending in the common bag behind), the mouth of the net being kept open by means of floats and weights; the dragging is done by manually operating the oars on either side of the boats.

Other common seine nets of Kerala are the *Thanguvala* and *Kollivala,* a special Kollivala called the *Pattenkollivala* being used with good results in recent decades. There is an *Ayilakollivala* for mackerel (corresponding to the *Ayilachalavala* the mackerel gillnet) and a *mathikollivala* for oil sardine (corresponding to the *mathichalavala* or sardine gillnet, the difference between the two being mainly in the mesh size. The general *pattenkollivala* is used for both mackerel and oil sardine as well as other fishes which come up in the net when the former two forms are not abundant.

Another interesting net which testifies to the ingenuity of the fisherfolk of our country is the *Dol* net of Bombay; this is a simple long conical bag of net of large meshes; this net is left suspended in water with the square or circular mouth opening oriented in such a way that when the water passes through the net the fish remain entangled and get caught, the rise and fall of the tide being utilised in the fishing. There are scores of other fishing nets and implements in use in different parts of the country, all these being modifications of a smaller number of basic patterns. A few rare types not conforming to the basic pattern also do exist here and there the Chinese dip net being one of these. This net is a squarish many-sided or just rectangular piece of net held in place by a framework almost in a "plane" level at margins but with a slight concavity to the centre; this is left entirely dipped in water for sometime and then lifted up by means of rope and pulley arrangement; whatever is collected on the net when it is lifted up is the catch of the single haul; obviously this does not sound a very efficient method and only the more sluggish fish are retained while others escape. But the method has its historical importance and is still perhaps used in some places.

In modern seafishing the kind of fishing very commonly done especially in shallow waters and wherever bottom fishes are involved is what is called trawling. The trawl is again an essentially bag type of net with a fish trapping cod-end and a pair of wings by means of which the net is attached and dragged by two mechanised vessels; the cod-end has valve-like contrivances so that the fish once taken in cannot come out and escape during the prolonged movement of the net in the water for over one to three hours. In order to keep the mouth of the net well open during the fishing activity, the wings are attached by ropes to a pair of planks called the *Otter boards*; as the water acts on the boards during the movement of the trawling vessel the net tends always to remain wide open. The indigenous *Paithuvala* mentioned above can in a simple way be compared to the trawl but with the difference that the *paithuvala* is only manually operated while the elaborately constructed trawlnet is motor-operated. While a single boat is used in otter-trawling with two otter boards as stated above, there is a kind of trawling called bulltrawling (Pareja trawling) or two-boat trawling where the net is bigger, has longer wings and has to be drawn by two boats each of which drags one of the two wings as in the case of the shore-seine or "Paithuvala". But in the case of bull-trawling and otter trawling the entire operation from setting net to the completion of harvesting is done on the sea itself, no dragging of the net to the shore being involved.

Two-boat trawling was initiated in India in or about 1949-50 off the Bombay and Gujarat coasts for the first time; this was done with the expertise of the

Japanese specialists with good results, and trawling (both experimental and commercial) is being continued even now with a large fleet of trawlers and with considerable profit. It has been found in fact, that in the same grounds the method of two-boat trawling can be much more profitable than the single boat otter-trawling. Like our boat-seine operations by indigenous methods off Kerala, the mechanised two-boat trawling seems to have been in vogue for several decades in some other seas, though in earlier years the publicity for this was not seen to the same extent as at present, and as over the past two or three decades. At present trawl-fishing is big business in Indian seas and it seems to have come to stay permanently as far as can be judged from the present conditions, which are very encouraging indeed.

The hook and line method of fishing, though most popular among sports fishermen is also used frequently in commercial fishing. Sharks, large catfishes, and tunas are generally caught by hooks and lines. In mechansed shark fishing the lines can be of long steel chains with numerous hooks arranged in a series and carrying plenty of chunks of flesh or fish as bait. Large sharks are killed by shooting before being hauled on board the vessels. Whales are caught after killing by harpooning. Whaling is however not a normal fishing activity for our country as the regular whaling area is far away in the Antarctic where we have had no fishing participation so far. Successful attempts at conquering part of the Antarctic area have been made by India recently from 1982 onwards, and we can perhaps look forward to some whaling by India also in future. Permanent Indian Stations are already working in the Antactica.

Fishing nets are normally made of cotton twines. Coir ropes and sometimes *"manila"* ropes of varying thickness and quality are used for jointing, for the head ropes, and also for various other complementary purposes in making the netting; jute material is also used in some places. In recent times nylon material is becoming popular owing to its durability; there also seems to be a belief that the nylon nets often bring in better catches.

In our seafisheries small scale fishing is done even now, mostly within the ten mile limit from the shore, this being about the maximum distance from shore that the fishermen can operate safely with indigenous equipment and gear; this is true of both the east and the west coast of the country.

(3) The Catches

Export of marine products from India have been steadily increasing in recent years. The Marine Products Export Development Authority of the Government of India (with Headquarters at Cochin) has been helping the industry in various ways for improving exports. During the year 1989 seafood worth over (600 crores) six hundred crores of rupees was exported from India. The United States of America, Japan and countries of the middle east and European west are the main purchasers of our seafish and fish products at present. The prawns and lobsters have proved to be good earners of hard currency, though the latter are still caught in relatively very small quantities only.

The most important sea fisheries of our country may be said to be formed by the Indian mackerel, *Rastrelliger kanagurta* and the oil sardine, *Sardinella longiceps*; these are most abundant off the Kerala and Karnataka coasts; these fishes are known as "Ayila" and "Mathi" respectively in the Malyalam language and "Bangude" and "Boothai" respectively in the Kannada language; the sardines and mackerel are called *"Choodai"* and *"Kumbala"* in Tamil. The oil sardine belongs to the family Clupeidae, the Clupeoids (the group including this family) contributing about a quarter of the fish produced annually in the Indian ocean; of this also most of the catch is obtained in the western Indian Ocean and only about 25% is produced in the Eastern Indian ocean. The Indian mackerel (*R. kanagurta*) forms about 9% on an average in the total marine fish catch of the Indian ocean and upto 20% of the Indian marine fish catch. This fish again is mostly caught in the western Indian coast only, though some landings are occasionally noticed elsewhere also.

During the five-year period 1972-76 the average quantities of the various categories of seafishes were as follows in the total average of 1325541 metric tons produced along the coasts of India (marine catch only, excluding the Islands). (Authority:- S.V.Bapat and Alexander Kurian (1981): Present status and role of small scale fisheries in India, in *Proceedings of Seminar on the role of small scale fisheries and coastal aquaculture in integrated rural development, 1978. CMFR Institute Bulletin No. 30-A, March 1981, p. 17)*

Table IV

Sl. No.	*Name of fish*	*Mean quantity in metric tons*	*Percentage*
1	2	3	4
1.	Elasmobranchs	56210	4.24
2.	Eels	5309	0.40
3.	Catfishes	131658	9.93
4.	*Chirocentrus*	10484	0.79
5.	a. Oil sardine	145367	10.97
5.	b. Other sardines	191567	14.45
6.	a. *Harpodon nehereus*	72756	5.50
6.	b. *Saurida* and *Saurus*	8158	0.61
7.	*Hemirhamphus* and *Belone*	1910	0.14
8.	Flying fish	2448	0.18
9.	Perches	25477	1.92
10.	Red Mullets	4871	0.38
11.	Polynemids	11169	0.84
12.	Sciaenids	82195	6.20
13.	Ribbonfish	55499	4.10
14.	Carangids	28654	2.16

(Contd.)

1	2	3	4
15.	*Leiognathus* and *Gazza*	44023	3.32
16.	*Lactarius*	10835	0.82
17.	Pomfrets	25170	1.90
18.	Mackerel	70571	5.32
19.	Seerfishes	20273	1.53
20.	Tunnies	10298	0.78
21.	*Sphyraena*	3076	0.23
22.	*Mugil*	3074	0.23
23.	*Bregmaceros*	2599	0.19
24.	Soles	13156	1.00
25.	a. Penaenid prawns	111418	8.86
25.	b. Non-Penaenid prawns	72713	5.49
25.	c. Lobsters	2641	0.20
25.	d. Crabs and other crustaceans	17239	1.30
26.	Cephalopods	5022	0.38
27.	Miscellaneous	73588	5.55
	Total:	1325541 metric tons	

The totals of fish production do vary from year to year. In the case of the economically important fishes these fluctuations or variations can sometimes be very great and are then a serious problem to the industry as well as the consumers. Fairly recent available annual catch figures as well as the figures of five-years 1976 to 1980 are given in table VI (authority: *Mar. Fish. Infor. Serv. T. & E. Ser.* No. 32, Oct. 1981). For the years 1956 to 1958 the following (table V) were the total production figures for seafish (all categories together) along Karnataka, Kerala and Maharashtra coasts as well as for the entire Indian coast (*CMFRI Bulletin* No. 13, 1969).

Table V

Estimated seafish landings in Karnataka, Kerala and Maharashtra coasts. (1956-68)

Years	*Total seafish Catch in Metric Tons*			
	(catch in m. tons) India	*Karnataka* catch %	*Kerala* catch %	*Maharashtra* catch %
1	2	3	4	5
1956	718779	20279/2.82	152213/21.18	244523/34.02
1957	875176	76090/8.69	309926/35.41	225670/25.79
1958	755994	80242/10.61	294655/38.98	147752/19.54
1959	584587	52825/9.04	191375/32.74	134617/23.03
1960	879681	110557/12.57	344605/39.17	127172/14.46
1961	683569	17248/2.52	267493/39.13	111839/16.36

1	2	3	4	5
1962	644244	43904/6.81	191421/29.71	123708/19.20
1963	655484	36514/5.57	202380/30.87	121337/18.51
1964	859582	100827/11.73	317727/36.96	129855/15.11
1965	832777	66102/7.94	334218/40.13	130552/15.68
1966	890311	62098/6.97	339402/38.12	133336/14.98
1967	869879	45272/5.20	359872/41.37	131823/15.15
1968	902772	82887/9.18	342065/37.89	121618/13.47
Average:	780987.3	61141.9/7.83	280488.6/35.91	144923.2/18.56

Table VI

Average estimated annual landings of marine fishes in India during 1976-80

Sl. No.		*Name of fish*	*5-year average in metric tons*	*Percentage*
1		2	3	4
1.		Elasmobranchs	57829.4	4.34
2.		Eels	9862.2	0.74
3.		Catfishes	45767.4	3.44
4.		*Chirocentrus*	11269.2	0.85
5.	a.	Oil sardine	151437.0	11.38
	b.	Lesser sardines	70793.2	5.32
	c.	*Hilsa ilisha*	8197.8	0.62
	d.	Other Hilsa	10746.4	0.81
	e.	*Anchoviella*	32685.6	2.46
	f.	*Thrissocles*	15572.6	1.17
6.	a.	*Harpodon nehereus*	103868.2	7.80
	b.	*Saurida* and *Saurus*	9422.2	0.71
7.		*Hemirhamphus* and *Belone*	1637.6	0.12
8.		Flying fish	1512.8	0.11
9.		Perches	34694.2	2.61
10.		Red mullets	3233.6	0.24
11.		Polynemids	7167.2	0.54
12.		Scianenids	93245	7.01
13.		Ribbonfish	63754.6	4.79
14.	a.	*Caranx*	25314.8	1.90
	b.	*Chorinemus*	3657.8	0.27
	c.	*Trachynotus*	91.2	0.01
	d.	Other Carangids	724.8	0.05
	e.	*Coryphaena*	199.8	0.02
	f.	*Elacate*	527.8	0.04
15.	a.	*Leiognathus*	45678..4	3.43
	b.	*Gazza*	302.8	0.02
16.		*Lactarius*	8560.2	0.64
17.		Pomfrets	38584	2.90

(Contd.)

1		2	3	4
18.		Mackerel	67931.8	5.10
19.		Seerfish	23518.0	1.77
20.		Tunnies	18637.2	1.40
21.		*Sphyraena*	2513.4	0.19
22.		*Mugil*	2187.6	0.16
23.		*Bregmaceros*	429.6	0.03
24.		Soles	12070.8	0.91
25.	a.	Penaeid prawns	113203.6	8.51
	b.	Non-Penaeid prawns	64809.6	4.87
	c.	Lobsters	1374	0.10
	d.	Crabs and other crustaceans	19991.8	1.50
26.		Cephalopods	12625.8	0.95
27.		Miscellaneous	93685.8	7.04
		Total:	1330892.2	

Table VII

Statewise data on coastline, fishing villages, active fishermen, indigenous crafts and fishlanding centres in the different maritime States of India

State	*Coastline (km)*	*Number of fishing villages*	*Number of active fishermen*	*Number of indigenous crafts*	*Number of fish landing centres*
1	2	3	4	5	6
Gujarat	1500	179	22518	4197	108
Maharashtra	600	299	41539	8288	173
Goa	110	40	4067	1118	40
Karnataka	270	145	21740	6248	95
Kerala	600	268	80898	21718	223
Tamil Nadu & Pondicherry	960	395	72105	32268	395
Andhra Pradesh	570	408	64592	25976	280
Orissa & West Bengal	1080	179	15076	6667	51
Total:	5690	1913	322535	106480	1365

(*Source:* S.V. Bapat and Alexander Kurian (1981): *Present status and role of small scale fisheries of India. Proceedings of the Seminar on the role of small-scale fisheries and coastal aquaculture in integrated rural development*, 6-9 December 1978, Madras. (Publication No. 30-A by CMFR Institute, Cochin, p. 14.)

The total fish production from the sea even when fish stocks are good, naturally depends to a great extent on the numbers of boats and nets in actual use according to the different categories and also upon the number of men working actively in the profession. According to the table VII it will be seen that according to the survey conducted by the Central Marine Fisheries Research Institute near the end of the last decade (i.e. the late seventies) the total number of fishing villages along the coast in the country was 1913 with 322535 active fishermen (going out to sea for fishing). In Karnataka there were 145 fishing villages and 21740 active marine fishermen; in Kerala and Maharashtra these numbers were 268 and 80898; and 299 and 41539 respectively. The annual average marine fish production amounted to 61141.9 m. tons during the 13-year period from 1956 to 1968 while it was 117400.2 m. tons for the five-year period 1976-80, along the Karnataka coast; for Kerala and the Maharashtra coasts the average annual marine fish production estimated for the above period of five years (1976-80) was 331895.0 and 273477.2 m. tons respectively. The total fish production of the entire world according to FAO statistics was 26,000,000 metric tons in 1956; there was a more or less steady increase in the production subsequently with some variations from year to year and country to country. The total sea fish production in India in 1983 was 1548475 m. tons while it was 1630678 m. tons in 1984 (*Mar. Fish. Infor. Serv. T. & E. Ser.*, No. 67, 1986). The total fish production of the entire world according to FAO statistics reached 73.5 million m. tons in the year 1976.

Although there are always slight ups and downs in the total annual production in individual fisheries in different regions, the overall totals may be said to have shown general trend towards a steady increase (with increased effort) in the world taken as a whole.

Considered from a long range point of view the progress of the sea fishery industry will naturally depend upon the progress of the research activity relating to the same; that is, the progress of the marine biology and other sciences of the sea and also the application of all these branches of knowledge in the practical exploitation of the resource. It seems to be the considered opinion of many experienced experts that the seafish production in India and also of the entire world can be increased considerably in the future years. One of the problems faced by the fishermen and the fishery industry is the unpredictability of the magnitude of individual fish catches in any given season fairly in advance; the onshore preparation for the utilisation of the landings will be easy if the above knowledge is available even very approximately a few weeks or even days earlier; cold storage, transport and marketing arrangements will be satisfactorily effective when the approximate requirements are known, as otherwise there can be wastage of both effort and money leading to loss in the related business. In the case of the Indian mackerel and the oil sardine for instance, the production in some years is high while in some other years it is very low; it is not yet possible to clearly and very definitely state also why exactly these fluctuations in the catches occur, and there is no predictable regularity noticed in these fluctuations. But the necessary supporting basic data on the various aspects of the fishes and their environment may become available with sustained researches and collected multi-faceted knowledge over a very large number of years only.

Table VIII
Estimated marine fish landings in India in metric tons during 1980

Sl. No.	Name of fish	W. Bengal	Orissa	Andhra Pradesh	Tamil Nadu	Pondi-cherry	Kerala	Karn-ataka
1	2	3	4	5	6	7	8	9
1.	Elasmobranchs	114	3772	4842	15442	435	6803	2910
2.	Eels	—	—	289	85	8	6	131
3.	Catfishes	723	2198	2338	4047	78	13936	5354
4.	Chirocentrus	276	1460	1123	2695	98	1002	171
5.	a. Oil sardine	—	—	—	320	—	69667	42727
	b. Lesser sardines	—	1891	13930	29940	2736	11017	4135
	c. *Hilsa ilisha*	644	5091	96	37	25	14	8
	d. Other *Hilsa*	20	46	1267	3084	53	21	25
	e. *Anchoviella*	—	270	6182	13126	287	7772	5621
	f. *Thrissocles*	194	333	7326	5048	387	2241	850
	g. Other Clupeids	674	2576	5486	1833	273	574	1088
6.	a. *Harpodon nehereus*	419	378	611	6	—	—	15
	b. *Saurida & Saurus*	—	189	931	1123	160	7080	508
7.	*Hemirhamphus & Belone*	—	46	97	749	26	361	180
8.	Flying fish	—	17	43	1106	3	-	55
9.	Perches	13	341	4639	6886	666	1784	1069
10.	Red Mullets	—	296	349	1079	150	1	38
11.	Polynemids	186	1126	1448	629	6	8	—
12.	Sciaenids	358	2864	9496	19547	320	6164	3500
13.	Ribbonfish	142	928	15646	7862	179	12937	1499
14.	a. *Caranx*	—	607	5981	5405	479	4399	4507
	b. *Chorinemus*	130	567	710	1111	2	145	67
	c. *Trachynotus*	—	—	—	38	2	—	—
	d. Other Carangids	—	—	97	188	—	59	232
	e. *Coryphaena*	—	—	3	141	—	138	—
	f. *Elacate*	—	—	19	148	—	19	3
15.	a. *Leiognathus*	34	704	3775	38153	681	4147	4671
	b. *Gazza*	—	3	56	84	—	1	42
16.	*Lactarius*	—	65	940	938	29	861	998
17.	Pomfrets	921	9072	2201	1306	188	907	696
18.	Mackerel	—	265	6203	7229	445	18474	19634
19.	Seerfish	234	1542	2970	7179	85	3763	1941
20.	Tunnies	—	34	419	4233	—	10611	952
21.	*Sphyraena*	—	8	88	932	55	330	84
22.	*Mugil*	—	1	27	577	49	151	39
23.	*Bregmaceros*	—	—	—	—	—	—	—
24.	Soles	3	69	573	2094	151	4394	782
25.	a. Penaeid prawns	152	1074	5660	9082	485	52633	3098
	b. Non-Penaeid prawns	48	30	4346	946	42	1742	128
	c. Lobsters	—	—	10	90	4	18	110
	d. Other crustceans	20	359	1413	6174	172	7286	2765
26.	Cephalopods	4	98	470	1472	40	4244	122
27.	Miscellaneous	788	1055	3913	15230	591	7803	4567
	Total:	6097	39375	116013	217394	9390	279543	115432

Table VIII (Contd.)
Estimated marine fish landings in India in metric tons during 1980 (contd.)

Sl.No.	*Name of fish*	*Goa*	*Maha-rashtra*	*Gujarat*	*Andamans*	*Laksha dweep*	*Private traw-lers*
1	2	3	4	5	6	7	8
1.	Elasmobranchs	894	7752	14558	56	284	—
2.	Eels	6	3154	8403	—	—	—
3.	Catfishes	1151	8653	5235	32	—	—
4.	Chirocentrus	124	2039	3792	25	—	—
5.	a. Oil sardine	2367	663	—	—	—	—
	b. Lesser sardines	1798	1363	—	243	—	—
	c. *Hilsa ilisha*	8	1017	56	—	—	—
	d. Other *Hilsa*	14	607	3965	25	—	—
	e. *Anchoviella*	249	78	—	99	—	—
	f. *Thrissocles*	779	1271	913	—	—	—
	g. Other Clupeids	302	16897	8538	29	—	—
6.	a. *Harpodon nehereus*	12	57393	36671	—	—	—
	b. *Saurida* and *Saurus*	199	1057	85	—	—	—
7.	*Hemirhamphus* and *Belone*	6	42	6	41	99	—
8.	Flying fish	2	—	—	—	29	—
9.	Perches	269	3712	2454	302	376	—
10.	Red Mullets	15	461	—	—	27	—
11.	Polynemids	10	1976	667	—	—	—
12.	Sciaenids	1530	13956	31625	—	—	—
13.	Ribbon fish	1089	11550	10858	—	—	—
14.	a. *Caranx*	884	1315	461	147	80	—
	b. *Chorinemus*	71	357	1022	—	—	—
	c. *Trachynotus*	—	—	—	—	—	—
	d. Other Carangids	—	369	—	—	—	—
	e. *Coryphaena*	—	20	—	—	—	—
	f. *Elacate*	190	—	—	—	—	—
15.	a. *Leiognathus*	1727	406	—	102	—	—
	b. *Gazza*	—	—	—	—	—	—
16.	*Lactarius*	614	450	2520	—	—	—
17.	Pomfrets	257	10081	12587	15	—	—
18.	Mackerel	2446	288	112	183	—	—
19.	Seerfish	735	3219	4180	117	21	—
20.	Tunnies	356	1674	277	55	1760	—
21.	*Sphyraena*	171	33	—	67	14	—
22.	*Mugil*	11	24	1034	117	—	—
23.	*Bregmaceros*	—	159	757	—	—	—
24.	Soles	1311	1797	2459	—	—	—
25.	a. Penaeid prawns	1853	23433	14481	54	—	32
	b. Non-Penaeid prawns	—	47309	4109	—	—	—
	c. Lobsters	18	225	204	—	—	—
	d. Other cristaceans	1933	297	4967	—	—	—
26.	Cephalopods	210	1191	3471	—	13	—
27.	Miscellaneous	879	5475	23027	94	206	2212
	Total:	24490	231763	203494	1803	2909	2244

4. Selected Important fisheries

We may now briefly consider the present state of knowledge of some of our more important seafisheries; these may be considered under the following four headings: (1) the oil sardine (2) the mackerel, (3) the flatfishes and (4) other fishes.

(a) The Oil sardine fishery of India

This fishery which is formed by the species *Sardinella longiceps* occupies the first place among the single species fisheries of our seas. This fishery showed a notorious decline in production during twenty years between the third and fifth decades of this century with the exception of the years 1922 to 1924, 1925-26 and 1933-34. After 1957 this fishery showed a recovery. The annual average catch of this fish for 1950-55 was 30200 m. tons but for the period 1956-60 the catch figure increased by four times, and the same increased to six times of the earlier figure during the years 1960-65; for 1965 the annual total catch of oil sardine in the Indian sea was 253302 m. tons out of the total seafish catch of 815120 m. tons, thus forming 31.69% of the marine total catch of the period. For 1976-80 the five-year average catch was 151437 m. tons; in 1980 it was 115744 m. tons, and in 1982 it was 205294 m. tons. These figures give an idea of the increase in the yields over the decades as well as the year to year fluctuations of individual fisheries mentioned earlier above. In 1983 the Indian oil sardine catch was 183706 m.t. and in 1984 it was 188832 m.t. (*Mar. Fish. Infor. Serv. T & E. Ser.*, 67, 1986)

Sardinella longiceps occurs in the seas of Arabia, Iran, Pakistan, Srilanka, Java and Bali islands and occurs abundantly during some months of the year in the shallow inshore waters of our west coast as an intensive fishery (off the coasts of Goa, Karnataka and Kerala); the fishery is well known in places like Mangalore and Malpe along the Karnataka coast and at Cannanore, Calicut, Tanur, Cochin and other areas off the Kerala coast. Though it can be stated that the oil sardine fishery extends upto Ratnagiri in the north and the coast of Srilanka in the south, the main commercial scale fishery is normally really restricted in abundance to the coasts of Kerala and Karnataka. On the east coast of India the fishery of the oil sardine is negligible and highly restricted, but there are other species of *Sardinella* (all these together being often known as the lesser sardines in commercial parlance) which form some minor fisheries in certain centres of both the east and west coasts of India. The more known among these include *Sardinella fimbriata, S. gibbosa, S. albella and S. jussieu;* the most important of these lesser sardines commercially is *S. fimbriata* and occurs in small quantities in many centres, sometimes in relative abundance also. *S. jussieu* and *S.albella* seem to be also caught in particular abundance on the south-east coast of India near Mandapam; Sekharan (1967) has worked on the lesser sardines of Mandapam.

Table IX
Estimated marine fish landings in India (in metric tonnes) during 1984 (Source: Mar. Fish. Infor. Serv. T.& E. Ser., 67, (1986), CMFRI, Cochin

Sl. No.	*Name of fish*	*W. Bengal*	*Orissa*	*Andhra Pradesh*	*Tamil Nadu*	*Pondi cherry*	*Kerala*	*Karn-ataka*
1	2	3	4	5	6	7	8	9
1.	Elasmobranchs	444	7681	10031	11925	425	7637	1601
2.	Eels	48	10	691	288	2	19	2
3.	Catfishes	2211	5993	5480	4197	68	11582	3722
4.	*Chirocentrus*	352	845	2034	2641	76	1372	501
5.	a. Oil sardine	0	539	0	1195	920	146893	36323
	b. Lesser sardines	24	3311	21077	26223	1155	6268	5445
	c. *Hilsa ilisha*	2101	1665	57	1127	0	198	15
	d. Other *Hilsa*	292	208	7195	5189	303	0	1
	e. Other clupeids	5316	4492	14665	28633	1918	48245	17795
6.	a. *Harpodon nehereus*	2166	303	1023	4	0	0	2
	b. *Saurida & Saurus*	0	132	1217	2184	130	6281	591
7.	*Hemirhamphus & Belone*	3	8	138	921	10	311	98
8.	flying fish	0	0	39	2157	451	3	1
9.	Perches	88	681	11121	11802	832	26873	2241
10.	Goat fishes	0	90	948	1758	43	75	1
11.	Polynemids	71	45	2130	475	23	453	13
12.	Sciaenids	2976	16903	8047	12707	741	9686	2005
13.	Ribbonfish	5654	1235	6457	11903	215	6464	733
14.	Carangids	101	940	7786	14671	1771	13672	7535
15.	Silver bellies	95	363	5035	39912	925	3911	3355
16.	*Lactarius*	0	18	594	834	0	1645	852
17.	Pomfrets	2674	2315	9869	1069	170	1613	1777
18.	Mackerel	32	413	6412	6540	951	11676	12338
19.	Seerfishes	249	1016	8072	6395	293	6244	4921
20.	Tunnies	31	31	866	2561	91	6169	1113
21.	Barracudas	0	25	158	1843	74	1109	12
22.	Mullets (*Mugil*)	11	1	93	423	3	69	42
23.	*Bregmaceros*	0	0	0	0	0	0	0
24.	Flatfishes (Soles)	38	327	597	1714	320	17806	6033
25.	a. Penaeid prawns	2304	2048	8787	15154	854	35529	5511
	b. Non-penaeid „	10735	21	1183	1453	63	738	0
	c. Lobsters	0	1	12	566	5	53	3
	d. Other crustaceans (crabs & stomatopods)	134	150	2774	9240	736	7560	10314
26.	Cephalopods	42	59	450	3694	37	5422	333
27.	Miscellaneous	1731	709	1956	20722	1336	7895	1768
	Total:	39923	46578	146694	252120	14941	393472	126996

Table IX (Contd.)
Estimated marine fish landings in India (in metric tonnes) during 1984
(Source: *Mar. Fish. Infor. Serv. T & E. Ser.*, 67, (1986), CMFRI, Cochin)

Sl. No.	*Name of fish*	*Goa*	*Maha-rashtra*	*Gujarat*	*Anda-mans*	*Laksha-dweep*	*Larger trawlers*	*Total*
1	2	3	4	5	6	7	8	9
1.	Elasmobranchs	713	11508	11050	339	287	116	57757
2.	Eels	28	2868	3049	0	0	107	7112
3.	Catfishes	1272	13418	9313	121	0	38	57415
4.	*Chirocentrus*	149	6203	2858	67	0	27	17125
5.	a. Oil sardine	1675	1287	0	0	0	0	188832
	b. Other sardines	2441	254	274	1122	0	0	67594
	c. *Hilsa ilisha*	0	556	72	106	0	0	5897
	d. Other *Hilsa*	1	2525	4844	0	0	0	20558
	e. Other clupeids	2300	20666	13985	208	0	0	158222
6.	a. *Harpodon nehereus*	0	58367	55887	0	0	0	117742
	b. *Saurida & Saurus* (lizard fishes)	736	2170	1620	0	0	129	15190
7.	*Hemirhamphus & Belone*	4	46	16	96	62	0	1713
8.	Flying fish	0	0	0	0	15	0	2666
9.	Perches	1640	5846	8678	864	205	390	71241
10.	Goatfishes	0	916	319	0	24	20	4194
11.	Polynemids	1	500	3846	41	0	5	7603
12.	Sciaenids	1677	22588	31877	30	0	695	109852
13.	Ribbonfish	479	11612	7972	0	0	361	53085
14.	Carangids	1991	5470	4271	505	45	55	58813
15.	Silver bellies	1669	1016	456	381	0	4	57122
16.	*Lactarius*	786	2475	7317	0	0	0	14521
17.	Pomfrets	386	19192	11295	37	0	32	49979
18.	Mackerel	2684	985	52	565	0	0	42648
19.	Seerfishes	366	4757	3856	320	59	103	36651
20.	Tunnies	150	2812	2002	215	4313	0	20354
21.	Barracudas	24	215	65	213	14	9	3761
22.	Mullets (*Mugil*)	93	59	2384	244	0	0	3422
23.	*Bregmaceros*	0	2650	0	0	0	0	2650
24.	Flatfishes (Soles)	2313	5078	6792	0	0	22	41040
25.	a. Peneid prawns	4853	43934	10848	199	0	30	130051
	b. Non-peneid "	0	39230	8536	2	0	0	61961
	c. Lobsters	6	963	1604	9	0	1	3223
	d. Other crustaceans (Crabs & Stomatopods)	5734	2628	18560	30	0	0	57560
26.	Cephalopods	408	7650	2312	0	14	543	20964
27.	Miscellaneous	3927	5841	14570	512	293	420	61680
	Total:	38505	306285	250590	6226	5331	3017	1630678

As one of the fisheries that have not yet been fully covered by any intensive mechanisation programme in harvesting, the catching of the oil sardine takes place mostly in the relatively shallow waters of the inshore fishing grounds within a depth of about 10 fathoms, often in much less depth of water. Of the various types of net used in the catching of this fish, a very efficient one seems to be the seine-net called the *Kollivala* of the Kerala coast operated as a two-boat unit; the *kollivala* used for the oil-sardine is slightly different in specifications from the one used for the mackerel, these two nets being separately named as the *mathikollivala* (for catching the oil-sardine) and the *ayilakollivala* (for catching the mackerel). A net called the *Pattenkollivala* (which is also another seine-net as mentioned earlier) has been becoming popular during past two or three decades and a general form of this net has yielded large quantities of oil-sardine as well as the mackerel in some years.

The oil-sardine fishery of the west coast starts a few weeks after the commencement of the monsoon, when large adult fish of 17-19 cm length groups enter the catches. In some years, juveniles less than a year old with total lengths of 4-5 cm length occur in large numbers even in the August-October period. As such juveniles continue to occur in the catches for several months continuously, it is assumed that the spawning season of the species may be a prolonged one lasting several months. As these sizes decrease in numbers in the catches subsequently, the slightly larger sizes of the 12-15 cm groups take their place and these contribute to the high production noticed during the months of September to January. Subsequently the fishery shows a gradual decline and the season comes to a close by April-May. When the oil-sardine occurs in great abundance, it is noticed that the temperature of the sea surface water shows a variation range of 27° to 29°C and a salinity range of from 33 parts per thousand to 35 parts per thousand of seawater.

There still seems to be some difference of opinion among the different workers in our country regarding the size to which the oil-sardine grows at different ages and also regarding the absolute age upto which the fish may be able to live. According to one opinion *S. longiceps* attains total lengths of 15, 17 and 19 cm respectively at the end of the first, second and third years of its life. If we consider the averages of different years the catches of the months of abundance are found to be made up mostly of the 0- and 1- year classes of the fish. In the oil-sardine of the 17-18 cm total length the gonads will be well developed and ripe. As such fish with well-developed gonads occur greatly during the November-December period it is suspected that the oil-sardine may be spawning twice in its life time. Upto now however, it was the general belief that the oil sardine may be spawning only once in its entire life. The eggs of this fish occur in the sea during the months of July to October-November. But the exact location of the spawning of the oil sardine has yet not been spotted successfully.

The main objective of fish research in programmes of fisheries investigations will be to understand their life-histories, their food habits, growth trends, movements, and such other phenomena relating to them. Of the various

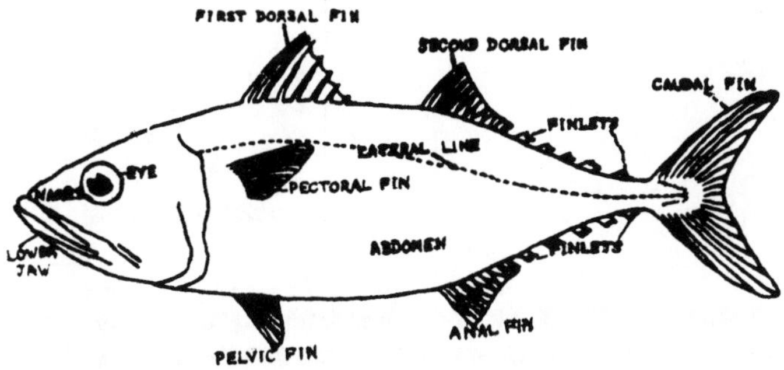

Fig. 41. (a). *Rastrelliger kanagurta* (Cuvier)—Left View

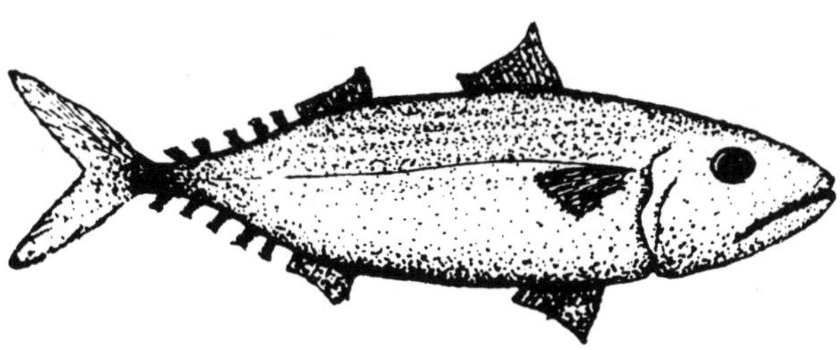

Fig. 41. (b). Rastrelliger kanagurta (adult)—(Right view)

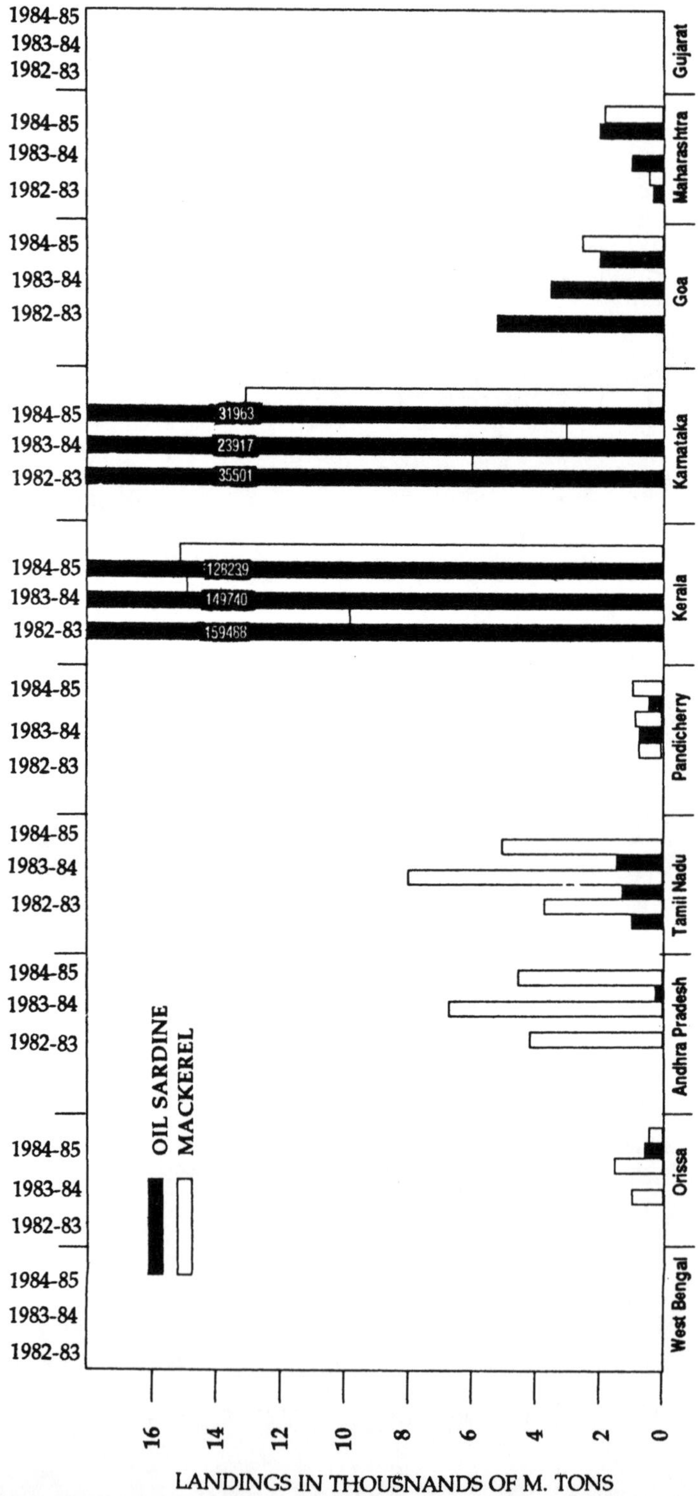

Fig. 42. **Statewise Total Estimated Landings of Oil Sardine and Mackerel in India during 1982-83, 1983-84 and 1984-85 (Source of data:** *Mar. Fish. Infor. Serv. T. & E. Ser.* **No. 67, (1986, CMFRI, Cochin).**

programmes the one aiming at studying the migrations of fish are the most difficult; tagging and releasing of tagged fish into the sea have been attempted recently in India for obtaining possible data for knowing growth and migration trends of the species. The recovered specimens among these give clues as to the growth attained by the fish during the interval between release and recapture, the distance and direction of migration and other relevant information.

(b) The Indian Mackerel, *Rastrelliger kanagurta*

Next in importance to the oil sardine is the mackerel fishery among the single species sea fisheries around India. The mackerel goes under the scientific name of *Rastrelliger kanagurta* and has the popular names of "Kumbala" in Tamil, "Ayila" in Malayalam and "Bangada" and "Bangude" respectively in Marathi and Kannada languages. It is called "Kanagurtulu" in Telulgu. In our annual seafish total production this fish is known to contribute from 3 to 20%; Noble (1985) has published some detailed calculations relating to the trend of mackerel landings in India in recent decades and years. The average all-India landings of this fish in 34 years from 1953 to 1983 was 66584 m. tons. The catches during 1950-53, 1957-60 (except 1959), 1963, 1969-73 and 1978-79 were higher than this and they were lower in other years. The average all-India annual value computed for the 19-year period 1965-83 was 67419 m. tons; the average annual production along the east coast only during this period was 8542 m. tons, while on the west coast it was 58877 m. tons. The 19-year average in the different maritime states were as follows (table X).

Table X
19 years average catch of mackerel in maritime states

Karnataka:	23478 m. tons	(=34.82% of all India total)
Kerala:	23094 m. tons	(=34.25% of all India total)
Goa:	9001 m. tons	(=13.35% of all India total)
Tamil Nadu & Pondicherry	5321 m. tons	(= 7.89% of all India total)
Maharashtra:	3293 m. tons	(= 4.89% of all India total)
Andhra Pradesh:	2688 m. tons	(= 3.99% of all India total)
Orissa:	436 m. tons	(= 0.65% of all India total)
Andaman and Nicobar Islands:	97 m. tons	(= 0.14% of all India total)
Gujarat:	11 m. tons	(= 0.02% of all India total)

West Bengal was reported to have caught 7 m. tons of mackerel in 1983 for the first time in the fishery history of that state; this was also the year when the east coast landings of the mackerel (=15680 m. tons) exceeded the west coast landings (=15503 m. tons) forming 50.3% and 49.7% respectively in the total along the two coasts. The total mackerel catch in India in 1983 was 31227 m.t. while it was 42648 m.t. in 1984. For the period 1965-83, Karnataka topped the list with 34.82% of the total mackerel production in the country. Kerala came next with 34.25%. The contribution of Goa in the all-India production was 13.35% and the states next in the order of importance were: Tamil Nadu–Pondicherry, Maharashtra, Andhra, Orissa, Andaman and Nicobar Islands, and Gujarat, these territories contributing 7.89%, 4.89%, 3.99%, 0.14%, and 0.02% of the total catches respectively (Noble, 1985).

The mackerel is a tasty and nutritious fish and fetches a much higher price per piece or unit weight than many other fishes. Of the total production of mackerel in India, it is interesting to note that as much as over 95% is contributed by the west coast alone, the east coast contribution being negligible. The main centres of the mackerel fishery on the west coast are Ratnagiri, Malwan, Karwar, Malpe, Cannanore, Calicut, Tanur, Cochin, Quilon, and Vizhingam. Just like the oil sardine, the mackerel also occurs in abundance in the shallow waters not far away from the shore during some months of the year. Off the coasts of Karnataka this fish is caught very near the shore at a depth of just a few metres. Considering the entire country it seems probable that the fish can be caught down to a depth of about 20 fathoms in certain regions; on very rare occasions the mackerel has occurred in the trawl catches but then in such small quantities that we have no sufficient justification to state that there is any deep sea fishery for the mackerel in our waters at present.

As in the case of the other fishes marine biologists have made some studies on the mackerel also from time to time in our country as elsewhere in the world. All researches have naturally to be circumscribed by the situations demanding the research in question and the facilities available for the same. If we have to understand the shoaling, movements and other habits of the fishes, we have to go on the sea and watch them at the spots of their natural occurrences, or rear them in tanks or aquaria of suitable sizes and note their activities under the experimental as well as natural conditions of existence. If we wish to know about the feeding habits of the fishes we have to feed them and see under artificial and natural conditions but as this is not always too easy, we have to cut open their stomachs and examine their contents under the microscopes. If we desire to know the growth rates of fishes we have to watch their growth in aquaria and in nature, but again, as this is not too easy and quick enough we have to take regular periodical samples at selected centres, note their lengths and weights regularly over a period of time, and from these data study the differences in the chosen parameters from place to place and from time to time so that we get a picture of the seasonal and yearly trends in these distributions. If we wish to know the times and rates of reproduction in these creatures (for estimating their productivity) we have to examine the gonads (i.e ovaries of females and testes of males) in the species from time to time and also note the periodicity in abundance (or scarcity) in the occurrence of postlarvae and juveniles including also the place and seasons of their occurrence. It is the total activity consisting of these several kinds of enquiry of a scientific nature involving laboratory as well as field work that goes under the name of fisheries research, in its biological division. But it must be emphasized that fisheries research does not end with mere work on fishes only; detailed understanding of the various aspects of the environment i.e. the physical, chemical and biological aspects of the sea water as well as the sea floor will be necessary, these aspects being equally if not often more important than the knowledge of the fishes themselves.

Some amount of scientific information about the mackerel is also available already as in the case of the oil sardine, this knowledge being the result of researches conducted at different centres during past few decades. Tagging experiments have also been taken up recently on a small experimental scale on both the mackerel and the oil sardine.

The mackerel fishery starts in our west coast sea in about October-November, a few weeks after the cessation of the south west monsoon. The mackerel occurring in the inshore catches of this period measures usually about 18-20 cm in total length; during the monsoon from July to September juveniles of the 10-12cm total length group also occur in small quantities; very small juveniles of 2-4 cm total length also occur in small numbers occasionally at certain centres, though not regularly. The markerel caught after November-December months is found to show a gradually increasing size trend in the catches from month to month and by about April-May a size of 22-24 cm is usually quite common. But by this time the fishery of the mackerel will have already shown a decline so that only small catches are noticed here and there along the coast. Though mackerel 24-27 cm long also do occur sometimes, their numbers are very small at most of the centres. Very rarely still larger sizes of 31-32 cm length have also been found at Mandapam on the east coast and at Vizhingam and Karwar on the west coast. The attempt to tell the age of the different sizes of this fish itself forms a distinct line of research. In any given catch of mackerel, either all or most of the fish may be of the same age—this generally being the rule in the mackerel and the oil sardine—or different age groups may be mixed up in the same catch. Some partial clue has been available from the scales of the mackerel for guessing the age of the specimens from which they are taken; these scales show in the bigger sizes of the fish some growth checks or rings in their structure when examined under the microscope (Fig. 45). Such rings have been so far noticed only in mackerel above 21-22 cm in total length. Research on this subject is still inconclusive but some provisional pointers seem to be indicated from the data collected so far. It would appear from these data that our west coast mackerel is about 12-16 cm at the end of the first year, 21-24 cm at the end of the second year, 25-27 cm at the end of the third year and about 28-29 cm at the end of the fourth year of life; the longevity may be 6-7years.

The food of the mackerel consists of mainly planktonic organisms; most important among the various groups of food organisms appear to be the copepods, which are minute creatures belonging to the Crustacea of the phylum Arthropoda. Some quantity of phytoplankton has also been occasionally noticed but the zooplankton is the more important constituent in the stomach contents of the mackerel, while in the oil sardine the phytoplankton is generally more important. Spawning in the mackerel takes place mainly during the months of June-September; in some years and in some places the spawning may start a little earlier or similarly continue a little longer than the normal period known for the centre or place. This delay or extension of the spawning season may sometimes be noticed by availability of evidence of spawning even in the months of December-January. The investigation of the spawning areas in the sea and the

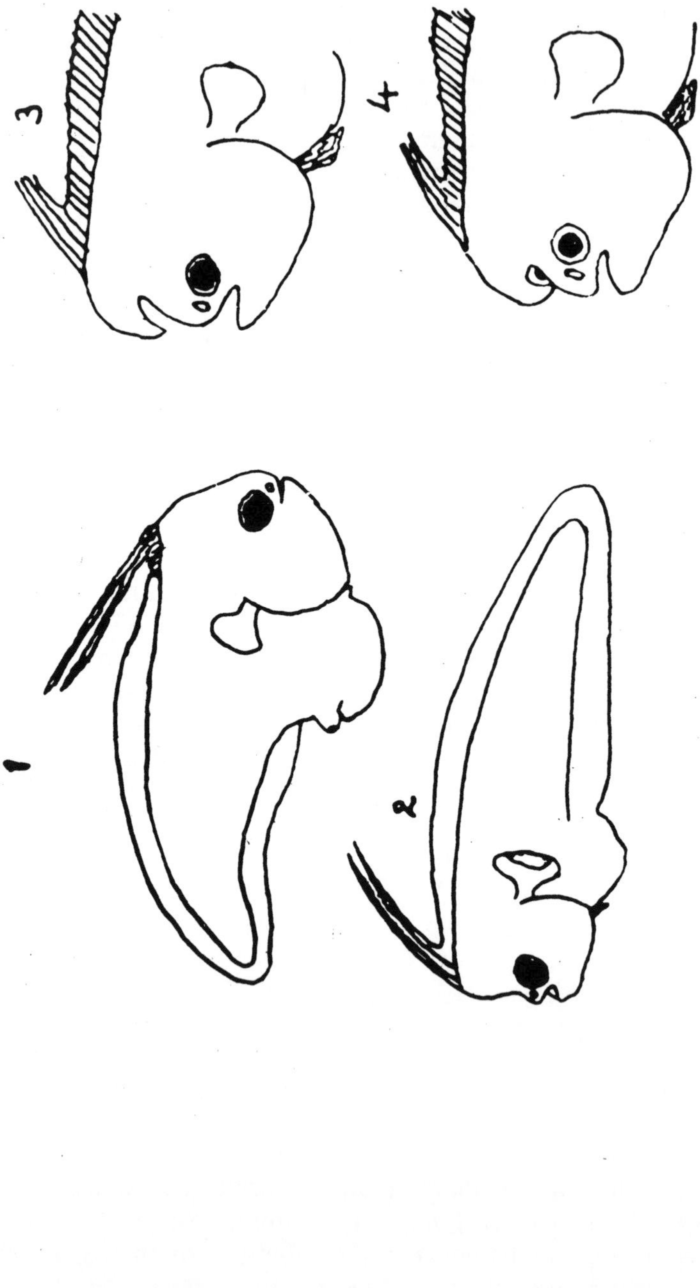

Fig. 43(a). Metamorphosis in *Cynoglossus semifasciatus* : (1). Symmetrical post-larva, 2. 8mm stage; (2). symmetrical post-larva, 4. 5mm stage. (3 to 5). stages in the shifting of the eye and the atrophy of the pectoral fin;

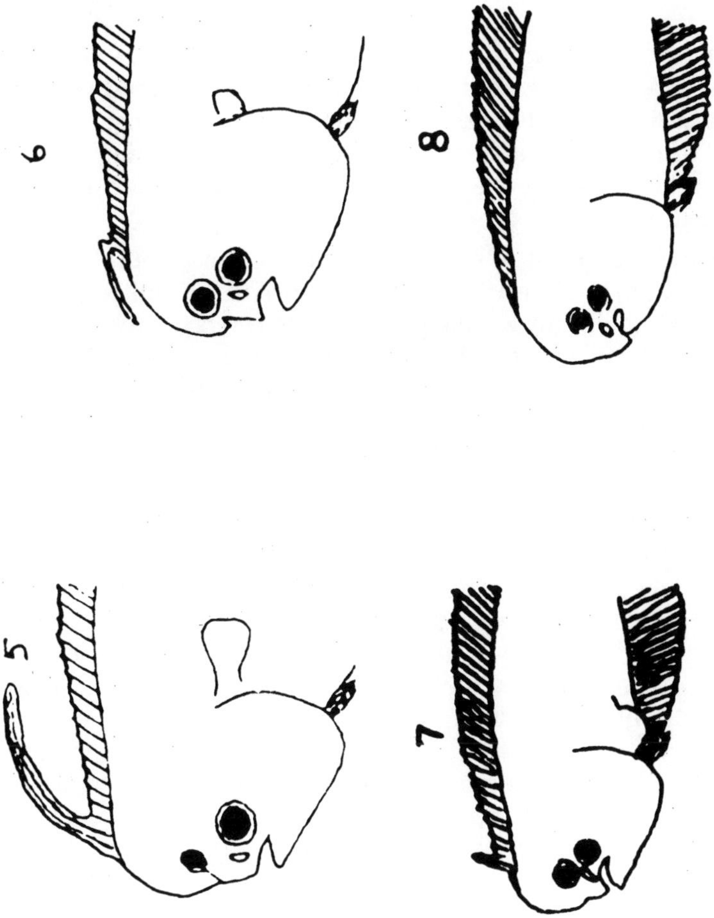

Fig. 43(b). (5 to 7) further stages in metamorphosis of C. semifasciatus ; (8) early juvenile (Based on Seshappa and Bhimachar, 1955)

density of the spawn and of the resulting larval and post-larval stages of the fishes is also another important field of research, though our progress in this line has also been somewhat slow in spite of the collection of much massive data. In the nineteen fifties and early sixties it was believed that spawning of the mackerel takes place only in one area near Vizhingam on the west coast, the juveniles migrating from here to other northern centres of the fishery. Recent trends of findings have been however contradictory to this and it is now seeming possible that spawning may be taking place at several places along the west coast. The spawning seasons are different for the east coast and the west coast and conditions of intermediate periodicity are found at Vizhingam where the spawning period of mackerel starts in March-April. The spawning is much later (i.e in June-September) along the west coast further north, and it is still earlier on the east coast.

As the same species that forms our mackerel fishery occurs around some of the far-eastern countries, the Indo-Pacific Fisheries Council of the United Nations Food and Agricultural Organisation is taking a keen interest in the research and development of the mackerel fishery in these areas. Even where our own species, i.e. *Rastrelliger kanagurta* is not the species forming the fishery, species of the same genus (*Rastrelliger)* contribute to the fishery and therefore a co-operative programme has proved very justified from the scientific point of view also.

Other genera and species of mackerel contribute to fisheries of varying magnitudes in the seas of Japan, west coast of United States of America and the North Sea. The chief mackerel of the Japan sea is *Scomber japonicus;* the Atlantic and North Sea mackerel is *Scomber scombrus*. In the genus *Rastrelliger* of our seas and of the far eastern seas, two species other than *kanagurta* are also known to occur, namely *R. neglectus and R. brachysoma* both largely outside India. According to some authorities the last two forms of mackerel belong to one and the same species. Our mainland form of mackerel (commonly referred to as the Indian mackerel) belongs to only one species namely, *kanagurta* as stated earlier. But the species *brachysoma* is said to occur in the sea around the Andamans and another species, *R. faughni* has been recently discovered at Madras (Gnanamuthu 1972); this form was not known along the Indian coast earlier.

(c) The Flatfishes

The flatfishes belonging to the Order Heterosomata possess some unique features and thus differ markedly from the other groups of the bony fishes. Because of certain peculiarities in their life history and in their body structure, these fishes are interesting not only from the fisheries point of view but also from a scientific angle. Among our marine flatfishes the Malabar sole *Cynoglossus semifasciatus* (Fig. 51) also referred to as *C. macrostomus* by some at present, is a well known fish of commercial importance. The body of the Malabar sole is elongated, flat and leaf-like with a body thickness of 1-2 cm or a little more in adults; there is an upper surface and lower surface to the body, and the upper surface usually has a number of dark half-bands on the skin across the body situated over a pale brown background colour of the skin. There are two longitudinal fins running along either margin of the body; one of these corresponds to the dorsal fin of other fishes while the other corresponds to the anal fin. The parts of the body, especially of the head region, display a marked asymmetry in their arrangement. Though there are differences between one group and other among the Heterosomata, all groups show this asymmetry as a common factor for the entire order. It is the opinion of evolutionary Ichthyologists that the flatfishes had a common origin with the perch-like fishes of the Acanthopterygii, and that after long process of evolution the asymmetry has established itself in the group. In the larval stages these flatfishes also possess the same bilateral symmetry as seen in the larval forms of other teleostean fishes; these larvae remain in upper layers of the sea water and as they metamorphose later on they descend downwards and settle down to a bottom life; as they do this they develop at the same time the asymmetry characteristic of the group, the migration of one of the eyes from its original site to the opposite side of the head being one of the essential happenings during this metamorphosis, this being characteristic of the entire group though the extent of migration and final position of the eye may vary from species to species. The interior part of skull also gets twisted to some extent during metamorphosis.

The eyes of flatfishes may both be either on the right side or on the left side of the head according to the different groups and species. The eyed side is however always the upper side and is the pigmented side, the lower side which is lying on the substratum having a pale whitish skin usually. In some groups there are species in which some specimens have the adult eyes on the right but others on the left; in *Psettodes erumei* about 50% of the specimens may have eyes on one side and the others on the opposite side. What exactly determines which side should finally be the eyed side is still an undetermined question but it is possible that the way the metamorphosing larvae settle down to bottom existence by what seems to be mere chance may decide how many in every hundred specimens are finally found to be of particular type of eye distribution.

Flatfishes occur in various parts of the world and some of them grow to a fairly large size and are of economic importance, being foodfishes useful to mankind. The plaice *Pleuronectes platessa* is a highly valued table fish in the countries around the North sea. In our country though *C. semifaciatus* is the most abundant fish other species such as *Psettodes erumei, Solea ovata* and *Pseudorhombus* spp form fisheries of some magnitude in some places; other species of *Cynoglossus* such as *C. dubius, C. macrolepidotus and C. bilineatus* also contribute some small quantities to the catches here and there, especially along the Kerala coast.

Though *C. semifasciatus* occurs on both the east and west coasts of India, it forms a commercial fishery of a good magnitude only on the west coast where again it is limited mainly to north Kerala and south Karnataka coasts. Normally this fishery starts immediately after the southwest monsoon and thereafter, within a couple of months the main fishery is mostly over. Spawning starts in October-November and by December itself small juveniles (of the new generation) measuring a few centimetres in length start appearing in the inshore catches. These juveniles grow up gradually while the recruitment of new batches of very small additional juveniles may also continue for several months. By about the following June the earlier batches of the juveniles reach about 8-10 cm total length and sometimes this size may be considerably exceeded also even some weeks earlier than this. With the commencement of the rainy season (in July) which brings about a fall in the salinity and temperature of the inshore waters, these fishes move over to slightly deeper waters where the conditions may be more stable and suitable; with the cessation of the monsoon rains and the return of normal climatic conditions in the inshore sea water by the end of September the soles enter the grounds in large numbers and the fishery starts again. At this period the soles come to the upper layers of the water in large aggregations which are called the shoals. Fig. 43 shows the metamorphosis and shifting of the eye in *C. semifasciatus* as observed at Calicut.

The age of the Malabar sole can be determined by the examination of its scales taken from the body. It is only in recent years that an attempt at using the scales of fishes in the tropical countries for age determination has met with some success. In the cold and temperate countries the seasonal differences in climate affect the growth trends of fishes and the differences between the winter and summer seasons leave an impression in the structure of the scales. During the spring and summer the scales of fishes in those countries grow regularly but they show a retardation and even a stoppage of the growth during winter, this happening leaving a ring-like appearance in the structure of the scales. As such marked differences of climate do not exist in our waters such growth rings are not expected in the fishes of our waters. But the west coast inshore sea is so much influenced by the monsoon rains and other associated conditions that this leaves a mark in the scales of the Malabar sole, this ring being designated the *monsoon ring* by its discoverers (Seshappa and Bhimachar 1951, 1954 & 1955 and Seshappa and Chakrapani 1983). The number of rings in the scales of a given fish will normally indicate the number of monsoons that the fish has passed in its life. Figs. 44, and 57 a, b, & c may be seen.

The soles of the species *C. semifasciatus* that occur in large quantities immediately after the monsoon every year are normally only one year old and show one growth ring in their scales; the dominant size at this stage is about 10-12 cm in total length. Two-year old soles measuring 14-16 cm in total length also occur in the catches in small numbers and these may be more numerous in some years but less numerous in other years; the scales of this size-group have two growth rings in them. Specimens of *C. semifasciatus* longer than 17-18 cm total length and of age more than three years (that is, with more than three true growth rings in the scales) have not been recorded in our waters so far. Individuals with three true growth rings in the scales have been noticed but only on very rare occasions. Abnormal "false rings" may however occur occasionally.

The Order Heterosomata includes five families and over 92 species in all, most of these being marine in habitat; *Psettodes erumei* and some of the larger forms of *Cynoglossus* such as *C. dubius* and *C. macrolepidotus* seem to occur mainly in the deeper waters of the inshore fishing grounds as noticed at Calicut (Kerala).

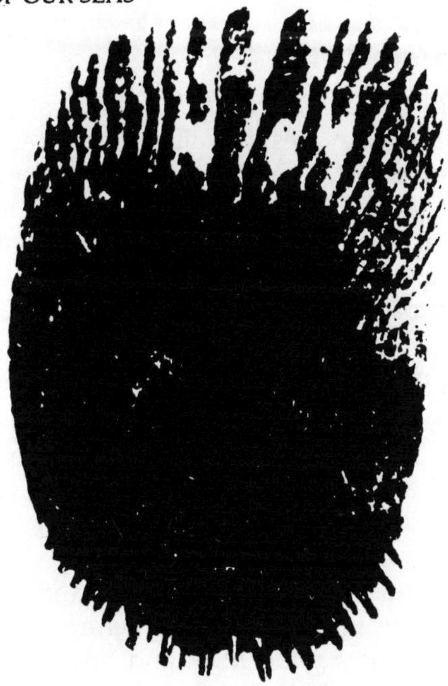

Fig. 44. Photograph of scale of *C. Semifasciatus* with two growth rings (Reproduced from Seshappa, 1951)

Fig. 45. Photograph of a scale of *R.kanagurta* with one growth ring (Reproduced from Seshappa, 1958)

(d) Other fishes

(i) *General:* In addition to the above, hundreds of species of other fishes occur in our seas at different places and during different periods of the year. Some of these are responsible for some fisheries of minor magnitude also and sometimes such fisheries may be quite considerable as in the case of the Bombay duck and the tunas. Only the more common of these will be considered here. The fact that these are grouped as *other fishes* here does not mean that any of these is of smaller importance as actually, after the oil sardine and the mackerel any of the following fish groups and species can occupy the next place of importance during certain years and at certain parts of the coast of India.

(ii) *The Tunas:* These belong to the Order Scombriformes the order including the mackerel also though in a different family. But the tunas grow to a size bigger than the mackerel and occur more in the upper layers of the oceanic and deep sea areas of the sea, that is, they are pelagic fishes occurring in the surface and mid-waters of the oceanic areas. A few examples of tunas are: *Euthynnus affinis, Neothynnus macropterus, Katsuwonus pelamis* and *Sarda orientalis.* Tunas occur plentifully at some of our coastal fishing centres in some seasons on both east and west coasts; There is a good hook and line fishery for the tunas near Tuticorin and off Mangalore and also around the Andamans. *K. pelamys* is the commonest tuna in the Laccadive or Lakshadweep islands where they are dried into solid sticks and marketed under the name of "masmeen"; this form of processing seems to preserve the fish for a very long time and masmeen is exported in large quantities from this area.

(iii) *Clupeoids other than the oil sardine:* The order and family of fishes that includes the oil sardine also include some more very important fishes. In the genus *Sardinella, S. longiceps* is the oil sardine; *S. fimbriata* supports a small fishery here and there on both our coasts, especially at Karwar, Malpe, Mangalore and other places down the coast. At Karwar where *S. longiceps* is negligible in the fishery, it is *S. fimbriata* that takes its place; this species is caught mostly in the *kairampani,* the small seine net there. Other species of Sardinella which go under the name of lesser sardines along with *S. fimbriata* are *S. albella, S. gibbosa* and *S. jussieu*; these form small catches with the miscellaneous catches as well as independently at different places. The famous *Hilsa* which can live both in the sea and the rivers and in fact ascends the rivers for spawing purposes, also belongs to this group. A few other examples of Clupeids are: *Anchoviella, Kowala, Thrissocles, Dussumieria,* and *Pellona*; these are all pelagic fishes of the inshore sea, but in shallow areas they are caught practically at all levels of the water.

The mackerel and the tunas belong to the Scombroid group as stated above; other examples of this group include the seer fishes of the genus *Cybium* or *Scomberomorus* of which there are three to four species all of which are good table fishes ranking in table value along with or sometimes even higher than the pomfret.

(iv) *The ribbonfishes:* (Fig. 52). These belong to the family Trichiuridae; there seem to be four distinct species of the ribbonfishes in our seas namely, *Trichiurus lepturus, Lepturacanthus savala, Eupleurogrammus intermedius* and *Eupleurogrammus muticus*; the first two of these species seem to be more important quantitatively in the fisheries. The ribbon fishes derive their name from the fact of their bodies being a very long, flattened and ribbon-like; they have also a silvery skin colour and possess strong canine teeth.

(v) *The Leiognathids:* These fishes of the family Leiognathidae are also known as the silver bellies because of the silvery colour of the abdomen and often of the rest of the body also. These are small fishes with as many as sixteen species in our seas belonging in all to four genera; the commonest of these seems to be *Leiognathus splendens,* the biology of which has been worked out in detail. The silver bellies occur in miscellaneous catches mixed with other small fishes.

(vi) *The "Bombil"or Bombay Duck:* This belongs to the species *Harpodon nehereus* which constitutes a dominant fishery on the west coast from Ratnagiri northwards off the Maharashtra and Gujarat coasts. Bapat (1970) has done considerable work on this fish and its fishery. Its production in India averaged between 1.36% and 17.93% of the total fish catch during the period 1950 to1965. During the period from 1971 to 1980 the average annual production was above 50,000 m. tons, the highest catches of the period, above 120,000 m.tons, being obtained in the years 1978 and 1979; the production of 1980 showed a decline of about 31,000 m. tons, as compared to 1979 , the 1980 figure being 95,505 m. tons. During 1981-82 the all-India total production of Bombay Duck was estimated to be 1,32,647 m. tons; in 1982-83 the corresponding figure was 90,422 m. tons there being thus again a fall in the catches. The total catch was 100950 m. tons in 1983 and it was 117742 m. tons in 1984.

The Bombil available in the markets measures from 6 cm to 27 cm in length; the fish attains maturity at the size of 21 cm length. The length of the fish has been determined to be 127 mm by the end of the first year and 217 mm by the end of the second year; the fish is ripe at the age of two years. A peculiarity in the occurrence of this fish is that it is seen in the northern part of our west coast and on a part of our east coast but none occurring in the intermediate region of the sea on either coast; the species is thus prominently absent in the south of the Arabian sea and on the west of peninsular India. A question that has arisen because of this discontinuity in the distribution of this fish is whether the two populations are really different races or variants of the same species or there are two distinct species in *Harpondon nehereus* occupying the two coasts with no occurrence in the intermediate region of the sea.

The Bombay Duck is a familiar sight off Bombay, especially at the Versova beach and the Sasson Dock where during many parts of the year the fish can be seen hung in large quantities for drying purposes on specially designed racks; the fish is sometimes dried spread out on the beach also. Compared to many other table fishes the Bombil is a cheap fish but may be called a poor man's fish and enjoys its own position in the market.

(vii) *The Jewfishes (Sciaenidae)*:- (Fig.38 & 48) This is an important family of perch- like fishes with the usual body-shape and piscine features common to what are sometimes called the "round fishes" (as distinct from the shell fishes and the flatfishes (the shell-fishes are of course Crustceans and Molluscs which

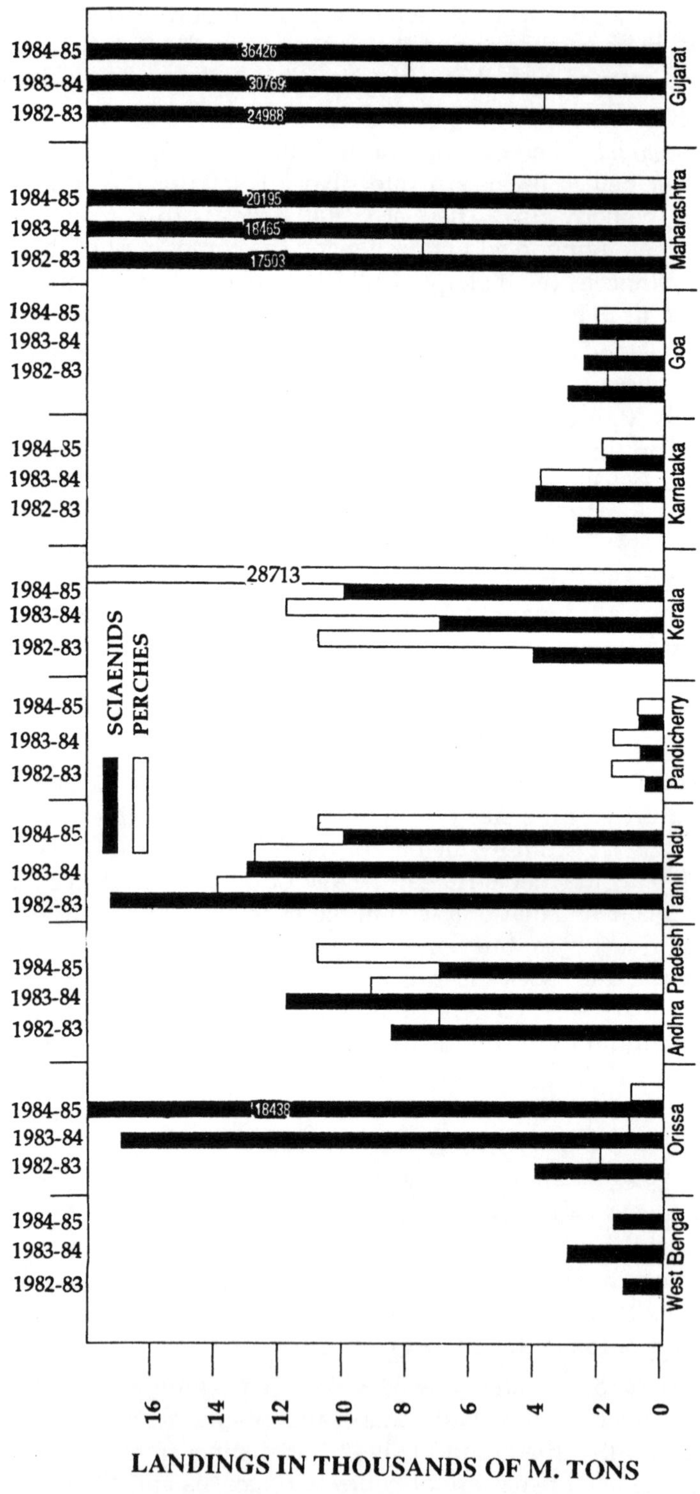

Fig. 46: Statewise Landings of Marine Perches and Sciaenids During 1982-83 to 1984-85 (Source of data: *Mar. Fish. Infor. Serv. T. & E. Ser.* No. No. 67, 1986, CMFRI, Cochin.

cannot be included with fishes in any consideration for scientific dealing but in commercial parlance they are often included along with the fishes as a separate category).

Some of the Sciaenids of the world are popularly called variously as "meagres", "grunters", "drums", "croakers", "Kingfishes", "weakfishes", etc; some of these names that suggest the sounds made by them are meaningful because of the peculiar grunting or croaking noises made by them under water. Various genera and species of the sciaenids which are also called generally as the jewfishes occur in various parts of the world and many of them are tropical also. No fossil remains of these fishes seem to have been found so far in any part of the world, and this suggests that the group may be in general a relatively recent family in its origin evolutionarily. There are five genera and over twenty species in our waters and a few of these are found living in freshwaters also but most of the species are marine in their habitat. (Lal Mohan (1972) has revised this group and recognises around 36 species in his new classification on basis of the air bladder) The family is represented on both our east and west coasts. Of all these the large-growing *Pseudosciaena diacanthus* (Fig. 48) may be placed as generally first in abundance and commercial importance. Called the "Ghol" in the Bombay region, this species is a most important constituent of the trawl catches in that part of the sea, especially in the offshore and deep sea waters. Ghols worth hundreds of thousands of rupees are also caught annually by indigenous gear in the Bombay region of the Arabian Sea. The trawl landings of fish from the Kathiawar region are dominated often by this fish. Other smaller jewfishes occurring with the Ghol are *Othlithes ruber, O. argenteus, Johnius belengeri,* and *species* of *Umbrina* (*=Sciaena*); as many of the other species are often less than 20cm in length the skippers of the larger mechanised vessels throw them back into the sea to save hold-space for the more valuable and often larger fishes. As the trawling durations and voyages are rather long they always exercise some choice in the matter of what fish to take back to the shore and what not to take. But in the inshore fishing practices even these small fishes (described as "Dhoma" together, meaning small jewfishes, or the trashfish category of jewfishes) are useful and marketed to be utilized either as food or for manure etc. These are numerous but their contribution is less than Ghol and Koth due to small size. The production of jewfishes, all species together, excluding the discarded part varied from 20442 m. tons to 32439 m. tons for the years 1957 to 1965; During the period 1971 to 1980, the best production year was 1975 with catches of the order of about 115000 m. tons; 1980 showed a decrease of 3,700 m.tons compared to 1979; there was actually a steady decline after 1977 when the total was in the order of only 10,000 m. tons. In 1981-82 however, the total catch of the sciaenids was 86,666 m. tons; in 1982-83 the total sciaenid catch was 82,170 m. tons, there being thus some reduction in the catches again. This reduction seems to have been due to decrease in the landings of Gujarat and Tamilnadu; the yield however seemed to have increased in the states of Orissa, Goa, Andhra Pradesh and Maharashtra. The Indian marine sciaenid catch in 1983 was 100755 m. tons and in 1984 it was 109852 m. tons this being a great improvement.

The Ghol by itself forms about 5% of the total fish marketed normally in the city of Bombay; the price of this fish is also more than several other kinds of fishes per unit weight, and the air bladders also fetch good price, being useful in the manufacture of "isinglass," which has strong adhesive properties and is useful in some industries, especially in foods and drinks.

Almost as important as the Ghol or even more important on a few occasions is the *"Koth" Otolithoides brunneus* (=*O. biauritus*) which grows longer than the Ghol though it is a bit slender comparatively. In shallow waters of a few metres depth only juvenile Koth of less than one year age are found dominating; the larger fish of the second and third year-classes are dominant in the deep sea trawlers off Bombay as a rule. The Koth lays a large number of eggs i.e. it has a very high fecundity and the spawning activity seems to occur continuously for several months; small juveniles occur now and then in the shallow waters near the shore. The scales of this fish also (like the Ghol) have age rings in them and these rings have indicated that the Koth may be able to live for five years or more. The food of this fish consists of other small fish and various animals.

The Ghol grows upto 120 cm or more in length and reaches first maturity at a length of 80-85 cm. The age rings in the scales seem to indicate that this fish can live upto 8-10 years of age.

Among the smaller sciaenids species of the genus *Otolithes* seem to occur practically at all places along our coasts; *O. argenteus* or the "silver bar fish" is the commonest among these. A number of other species of small jewfishes are together (in the mixed commercial catches) marketed as "Dhoma" at Bombay and other places as mentioned above.

(viii) The Elasmobranchs (Fig. 49 &50): It is estimated that the Elasmobranchs or cartilaginous fishes form about 6% of the total sea fish on an average. Shark fishing is in vogue along both our coasts, this fishery coming under the commercial activity of "deep sea fishing". The Governments of Madras, Bombay and to some extent Bengal were giving considerable encouragement for shark fisheries even during the earlier years of the century before Indian Independence. Sharks and skates are usually caught by Hook and Line; they may also get caught to varying degrees in other nets, the larger sharks however requiring large mechanised boats for their capture. Large sharks getting caught unawares in trawl nets often cause great damage by tearing the nets. Many forms of small sharks and rays occur in the shallow waters and form part of the miscellaneous catches of these depth regions.

For 1968 the Elasmobranch fishes formed 31028 m. tons out of the all-fish total catch of 902948 m. tons; in 1980 the total in this category was 57,862 m. tons; in the 1971-80 period the best years were 1974 and 1975 when the catches were in the order of about 67,000 m. tons, 1977 and 1978 being next in order with about 65,000 m. tons; the lowest catch during the above period was in 1971 with the total of around 42,000 m. tons only. In 1982 the total Elasmobranch catch was estimated to be 64,316 m. tons, of which sharks constituted 39,764 m.tons, skates formed 3,345 m. tons and the rays were 21,207 m.tons.

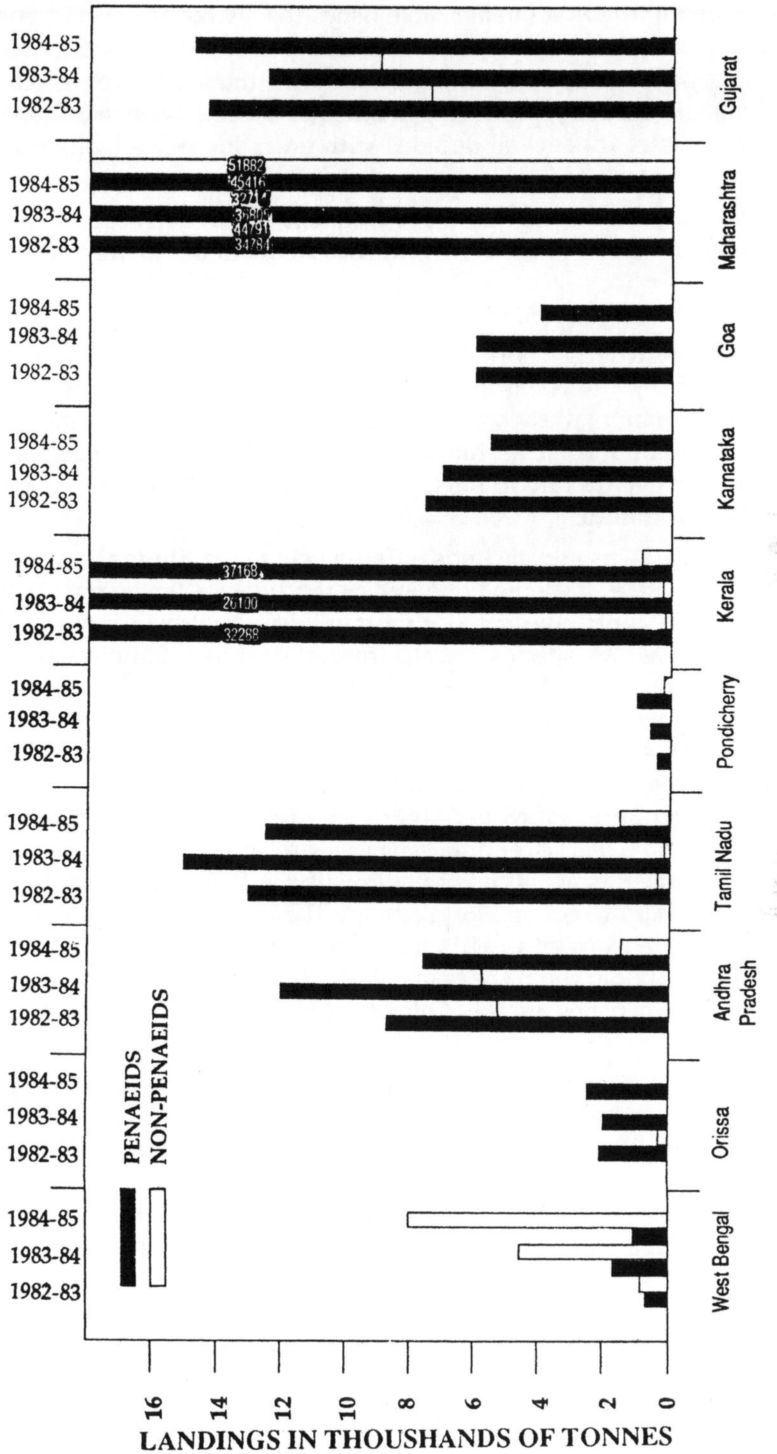

Fig. 47: Statewise All India Landings of Marine Penaeid and Non-Penaeid Prawns during 1982-83 to 1984-85 (Source of data: *Mar. Fish. Infor. Serv. T. & E. Ser.* No. 67, 1986, CMFRI.

Though not among the first quality table fishes the sharks are also in good demand both as fillets and as whole fishes (in the case of the small ones); shark fins are said to be good for soup and rich in certain proteins; shark liver is rich in oil; the shark liver oil industry producing medicinal oil has been working in Kerala and Bombay States for several decades with great success; a cattle brand of shark liver oil is also marketed.

Of the skates, *Rhynchobatus djeddensis* is frequent in our offshore water in the larger sizes; several species of small rays are also frequent in our waters.

(e) Prawns and Crabs (Figs. 54 & 55)

In addition to the true fishes discussed above, various forms of prawns, lobsters and crabs among the Arthropods (Crustacea), lamellibranchs (the true shellfishes), and other molluscs (such as the *Sepia, Octopus* and *Loligo* of the class Cephalopoda are also important as human food. There are many species of prawns of which over twenty important species contribute to dominant marine fisheries in variable magnitudes.

The commercial prawns and shrimps of India belong mainly to three major families namely, Penaeidae, Palaemonidae and Sergestidae; a few forms of the family Pandalidae also contribute some important deep water forms. George (1969) lists as many as 73 species in all from the above families as the commercially important forms. Mohamed (1969) states that there are 28 valid species in the genus *Penaeus* itself, eight of these being represented in Indian waters and also being commercially important: "Most of the species belonging to this genus grow to large sizes and they support commercial fisheries in many parts of the world, accounting for over 50% of the total world production of prawns and shrimps." *P. indicus* and *P. monodon* are among the commonest species of *Penaeus*. The genus *Metapenaeus* has 24 species of which 10 have been recorded in Indian waters (George 1969). Among the commonest of these are *Metapenaeus dobsoni*, M. *monoceros, M. affinis*, and *M. dobsoni; Parapenaeopsis stylifera* is also found abundantly in the boat-seine fish collections along with the Malabar sole and other fishes in the Calicut inshore waters (Bhimachar & Venkataraman 1953).

Mode of reproduction, development and metamorphosis are interesting in the crustaceans. The sexes are separate and there is sexual dimorphism, some of the appendages becoming modified in both sexes for reproductory purposes. Fertilisation is external; the eggs give rise to a freeswimming *Nauplius* larva which passes by moulting, growth and change through several further stages (sometimes over a dozen as in the case of *Metapenaeus dobsoni*), the chief of which are the *Protozoea, Zoea, Mysis*, and *Megalopa* stages (with substages), the last being characteristic of the pre-adult crabs. While in the prawns and shrimps the earlier larval stages are usually clear, in the crabs they are passed through in the egg itself. In the lobsters the first larval stage is a modified *Mysis* giving rise by moulting to a flat broad leaflike stage called the *Phyllosoma* which has long biramous appendages and metamorphoses into the adult.

In species where free larval stages are not visible in life, these stages are passed through within the egg itself and there seems to be evolutionary difference between this kind of reproductional history and the cases where all the larval stages are existing in the life history as freeswimming stages. Parker and Haswell (1947, p. I-451) however mention a case of the same species (*Leptodora* sp.) of the Order Cladocera where in summer a freeswimming stage is absent but in winter forms a freeswimming larva is present.

Larval curstaceans of various groups including the prawns and crabs and occasionally lobsters also form important constituents of the plankton both in the sea and estuaries in certain seasons of the year. Most prawns ascend into the rivers as juveniles while *Parapenaeopsis stylifera* is known to spend all its life in the sea only.

Among the marine species *Penaeus monodon* can grow upto a length of about 25 cm while *Macrobranchium rosenbergi* (the backwater prawn) can grow even bigger upto 31 cm in length. Over 80,000 m. tons of prawns are produced annually on an average in the Indian seas, forming about 10% of the marine fish production, *sensu lato*. About three-quarters of the country's production of the sea prawns is landed from the sea along the Maharashtra and Kerala coasts only. During the year 1980 the total catch of prawns from Indian seas was 1,70,137 m. tons of which 1,12,037 m. tons was constituted by the Penaeid prawns and the rest by the non-Penaeid prawns; including the lobsters and other crustaceas (crabs etc) the total was 1,96,802 m. tons. During the period 1971-80 the best catch of Penaeid prawns was obtained in 1975, the next best years being 1973 and 1978 in that order, the catch being slightly less in 1978. The 1980 catch of this group showed a marginal decline over 1979 by about 1,600 m.tons, the catch figure for 1979 being 1,13,665 m. tons. In 1981-82 the total crustacean catch from the Indian seas was estimated as 1,65,024 m. tons of which 87,112m.tons consisted of Penaeid prawns, 50,964 m. tons consisted of non-Penaeid prawns and the rest was constituted by lobsters, crabs and Stomatopods. In 1982-83 the Penaeid prawns formed 1,17,467 m. tons, and the non-Penaeid prawns formed 56,094 m. tons; lobsters formed 1,389 m. tons and crabs and Stomatopods formed 21,930 m. tons and 24,930 m. tons respectively in the total crustacean catch of 220,860 m. tons which itself formed 15.6% of the all-fish total catch of the year. The total prawn catch was 166953 m. tons in 1983 and 192012 m. tons in 1984.

Prawns are caught in various types of bagnets and their modifications; in recent years the prawn fishery has developed considerably and a number of private fishing and trading companies have been doing business in this commodity—catching, processing as well as trading in it—with good and encouraging results. The larger sizes among the prawns caught are usually exported to the U.S.A, Japan and other countries.

Along the south west coast of India the fishery may be said to last from about November to May usually, starting a few weeks after the cessation of the south west monsoon and continuing till the commencement of the next one. Cochin is an important centre for the prawn trade in Kerala.

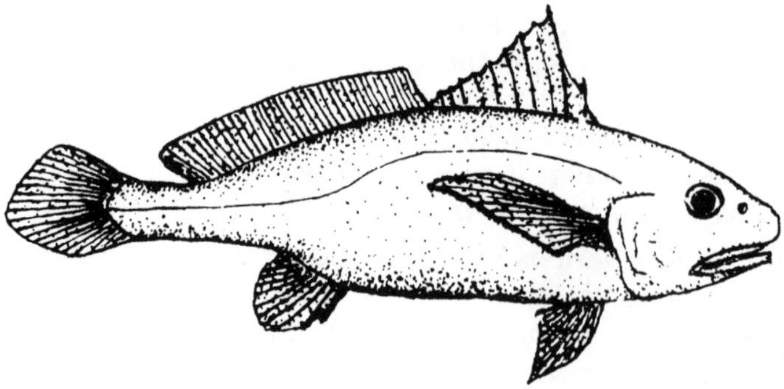

Fig. 48. The Ghol (Family Sciaenidae)

Fig. 49. The Dogfish (*Scoliodon sorrakowah*)

Fig. 50. The Sawfish (*Pristis Sp.*)

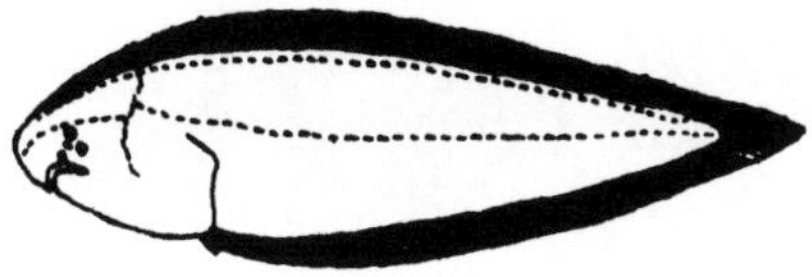

51

52

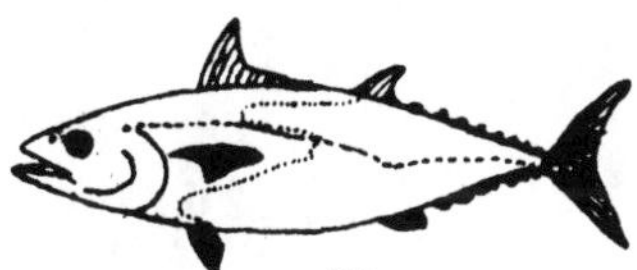

53

54

55

56

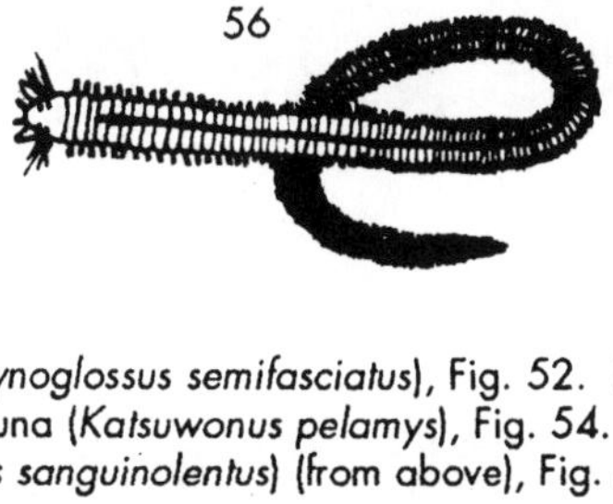

Fig. 51. The Malabar Sole (*Cynoglossus semifasciatus*), Fig. 52. Ribbonfish (*Trichiurus savala*) Fig. 53. The Skipjack Tuna (*Katsuwonus pelamys*), Fig. 54. A Prawn (Side View) Fig. 55. A Crab ((*Neptunus sanguinolentus*) (from above), Fig. 56. A Polychaete Worm.

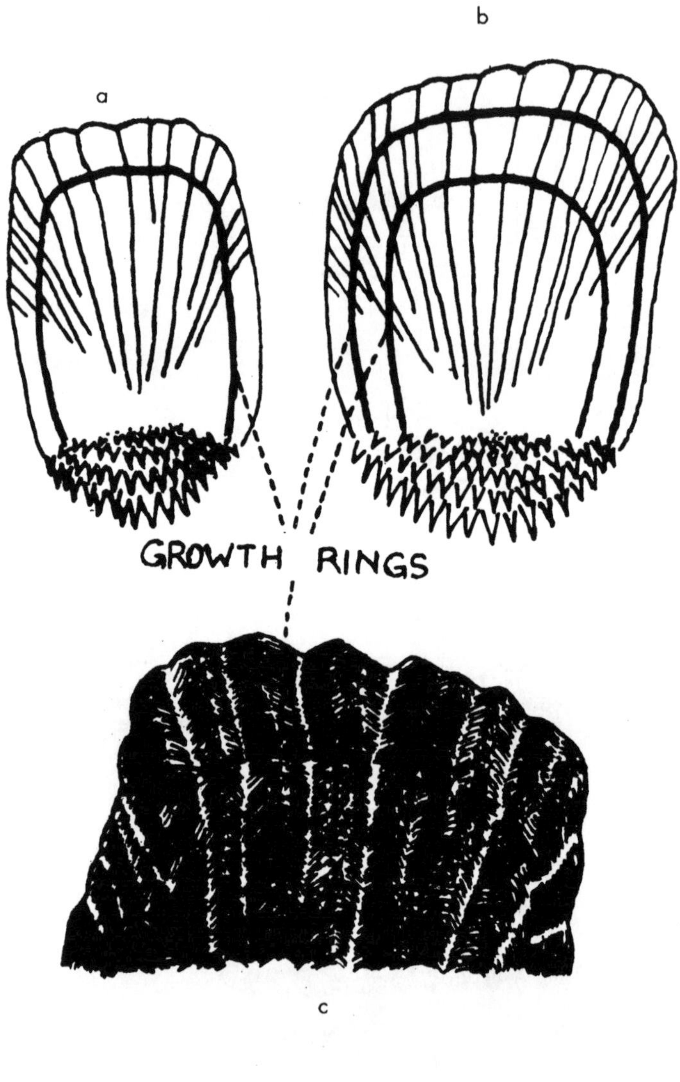

Fig. 57(a, b and c). Enlarged Drawings of Scales of Malabar Sole (*C. Semifasciatus*) Showing Structure and Position of Growth Rings) (a). scale with one growth ring; (b). scale with two growth rings; (c). part of scale enlarged to show one ring and structure of surrounding part of scale. (Re-drawn from Seshappa and Bhimachar, 1955)

Many species of prawns spawn in the sea itself, the juveniles entering the backwaters and rivers also. Some species of deep sea prawns have also been discovered recently in our waters off the south west coast; examples of these are *Aristius semidentatus* and *Parapandalus spinipes*. It may be anticipated that some new finds also may prove to be good fishery resources.

The lobsters are responsible for some fisheries in some centres along our coasts. Though these are caught so far only in rather small quantities they are valued higher as food and also fetch a much higher price per unit weight.

Crabs are also useful as human food and belong to the same major group (Crustacea) as the prawns. Species such as *Portunus sanguinolentus* are common on the shore as well as in the shallow sea. There are various other common species, some occurring in the deeper waters of the sea. Jones (1967) lists the following species as the main ones forming the crab resources of the Indian seas: *Scylla serrata, Portunus pelagicus, P. sanguinolentus, Varuna literata, Paratelphusa spinigera, P. jaquemontil,* and *P. hydrodromus*. The last three among these are freshwater forms and *V. literata* is a non-swimming small crab of the gangetic delta. *S. serrata* is a widely distributed brackish water crab adapted to live for long periods in freshwater also and growing upto 15-20 cm across the carapace. This is our largest and commonest food crab and occurs throughout the year. *P. pelagicus* is another common form of crab and grows to about 15 cm across the carapace. It occurs in the inshore sea and tolerates brackish water also; and it occurs also in the Chilka lake in almost freshwater. *P. sanguinolentus* is common all over the country and is caught along with *P. pelagicus* in the shore-seines, castnets, trawls etc.

In addition to prawns and crabs must be mentioned the lobsters which are of some commercial importance. There are four species; *Panulirus polyphagus, P. ornatus, P. homarus* and *Puerulus sewelli*. The first three of these grow to beyond 30 cm in length, *P. polyphagus* growing the largest among them (35-37 cm).

It may be stated here about the crabs that only the ones growing to a relatively large size among them are really of food value and several of the smaller crabs are of no use at all; many such species of shore crabs occur on rocky shores. The hermit-crabs are of course no use at all as human food.

(f) The Molluscan Fisheries

As already mentioned many of the molluscs are popularly mentioned as shell fishes. Among the Pelecypoda (Lamellibranchiata or Bivalvia) the common examples of fisheries importance are the clams, mussels and oysters. Among the Gastropods there are some chanks and chank-like animals (e.g. *Trochus*) that are important; *Trochus* grows to a large size and is caught and exported in considerable quantities from the Andamans region to foreign countries. The shell of this gastropod is used in the manufacture of buttons and some handicraft articles, several other species of gastropods being similarly useful so far as the shell is concerned.

Among the clams species of *Meretrix* are common in our backwaters and estuaries, the juveniles of these species and sometimes adults also being capable of living in the shore and the shallow sea. Clam meat is eaten and clam shells are used in lime-making and other industries. While the clam is poor man's food the oyster is a delicacy for the more sophisticated people of gastonomic temperaments. Edible oysters of one species or the other seem to occur in all parts of the world. The rock oyster *Ostrea cucullata* does not seem to enjoy much popularity as food though it is a very interesting form for study, being an intertidal species growing on rocks. The pearl oyster *Pinctada fucata* is world famous for the precious pearls of commerce which are globules of a shining nacreous substance secreted around foreign particles that enter between the mantle and the body of the animal. Several other bivalve shells other than clams are also often used for lime burning. Mussels are a source of human food in many places and in the past two decades large scale culture of mussels is being undertaken in some places along the coast in the shore and shallow sea.

We have already stated that there are the Octopoda and the Decapoda among the Cephalopods, the former being exemplified by the devilfish *Octopus* and the latter by the cuttlefish *Sepia* and another form, the *Loligo*. These last two also form some fisheries at different centres along our coasts.

The sacred chank used in our temples and in the domestic *Pooja* in the Hindu families is the shell of a gastropod snail called *Xancus pyrum*. The shells of this animal may be whorled rightward or leftward (i.e. dextral or sinistral), the leftward or sinistral shells being rare and highly valued, costing sometimes thousands of rupees *each*. Chank and pearl fisheries have been in practice along the east coast of Tamilnadu in the Gulf of Mannar for several decades and these yield a good revenue to the Government, this region of the sea falling within the jurisdiction of the Tamilnadu Government. There is a pearl oyster fishery off the Kathiawar coast and in the Persian Gulf there has been a famous pearl oyster fishery for a very long time.

Chapter X

Some of the Methods and Equipment for Research

(a) Scope of Research

For detailed researches on the plants and animals of the sea and for obtaining whatever knowledge can be obtained about them, all the necessary means and facilities are naturally required. Generally the nearer such laboratories are to the sea, the better it will be as that will make it easy for the workers to visit the sea frequently and obtain samples of various kinds of materials that will be of interest from time to time for examination and research; more over, many kinds of observations may have to be made on the spot itself, observations on the various organisms in their own surroundings in nature in the living condition; observations on living organisms often can give a picture quite different from what is gathered from dead or preserved organisms, each of these complementing the other for making a complete picture. Various kinds of equipment are necessary for the collection of different kinds of samples from the sea. All facilities will be of a limited nature everywhere at the beginning much depending on the availability of funds and the interests of the actual workers on the spot. Each line of work has to be considered as a special discipline needing its own allotment of funds for a library, instruments, accommodation and equipment. Naturally, availability of allround facilities will take its own time. The objective of every line of ecological enquiry in the sea will ultimately be to relate the results to the overall picture of life in the sea, its biology and full cycle. However limited the problem under investigation is at any given time, its ultimate value has to be judged in the way its answers fit into and complete the mosaic picture aimed at for the entire sea or part of the sea, its biology and fishery cycle; the utilization of the sea and its resources for man is dependent on this mosaic picture which has to be as complete and reliable as possible; in any branch of scientific enquiry the fundamental or basic questions to be answered by investigation will be "what", "how" and "whither"; quantitative answers to the questions will be the ultimate objective though it may not always be possible to provide clearly quantitative answers to the questions of most natural phenomena in the beginning.

The advantages to mankind from the sea are two-fold: (1) Firstly, from ancient times, the sea has been providing the route for communication, travel, trade and defence among the continents and among the maritime countries; (2)

secondly, the produce of the sea in its various ways is being utilized by mankind all through human history; this includes both living and non-living as well as derived resources.

The knowledge of the depths of the different areas of the sea, the mapping of the rocks occurring in different regions, knowledge of the tides and currents in different seas of the different parts of the world, the knowledge of predictable storms and hurricanes in certain areas of the sea, knowledge of areas that may be regularly haunted by sharks and other harmful animals, all these will be helpful in providing better security to the people concerned with the use of the sea as a medium of transport and commerce, let alone the most important need of national and international defence. As the depths of the sea in the different parts of the world have been already mapped so far as the shipping routes are concerned, there is no inconvenience to the ships sailing in the high seas today. In the old ages no dependable knowledge of the navigable areas of the sea was openly available to all concerned with such navigation; much difficulty was experienced in the past under such circumstances by the crew and passengers of ships sailing in the sea. Added to the difficulty about the accurate knowledge of the navigable and non-navigable areas in the sea was also the threat of pirates and open sea robbers adventuring on their own to make their fortunes by the practice of their desperate and inhuman profession.

The utilization of the sea for the benefits of man is also not just a recent happening; fishing in the sea is older than the days of even the Bible. As and when new and useful things became known from the sea, as and when new uses to things came to light, as new routes for sea navigation opened out for general use, and as facilities of mechanisation and technical know-how increased more and more, attempts were made more and more by man for utilizing the resources of the sea to a greater and greater extent; this trend has been evident at every stage of human progress. Under the present circumstances, leaving aside the short term or immediate objectives, the real long-term objective of sea fisheries research may be said to be to find out, in keeping with the increasing knowledge, new useful materials, their places of origin as resources, and to find out new uses for items already known and in use; ultimately the objective really would be to find ways and means of getting out the maximum from the sea but in such a way that the future supplies of these resources will not be adversely affected; in other words, the optimum utilisation is to be made of all the sea resources by the scientific application of all the branches of knowledge on the sea and on its living as well as non-living resources.

(b) Methods

A few of the methods used in marine research in its elementary aspects and in the collection of the samples for the research may be detailed very briefly in the following paragraphs:

(1) Sounding the depth of the sea

This will be among the first things to be done in any marine investigation, no samples being of any use unless the depth of the sampling area is known. One of the simplest methods in use from ancient times for finding the depth of the sea in any given place would be to tie a standard weight to the end of a piece of good string or rope and let it down vertically into the sea. It was customary in the old days to use lead weights of about 12-14 lb. heaviness each smeared with some thick oil for the purpose. The oily surface of the lead piece would come up when hauled, with sand or mud particles sticking to it, the nature of these particles indicating the nature of the sea bottom sampled; the length of the rope given out for the weight to reach the bottom would give the depth of the water at the spot sampled; knots tied in the rope at measured intervals of say, a meter each would make it very easy to assess the length of rope given out in the sounding process. While working in shallow waters, even ordinary stone weights could be easily utilised for the purpose in the absence of standard weights in any emergency; the advantage of using fairly heavy weights is in the fact that the rope will descend straight, as otherwise it may go down at an angle only, especially in deep water, this making the depth assessment subject to a slight error.

In modern shipping and navigational practices there are various electronic devices for direct reading of the depths where the water samples are taken; the nature of the bottom is also to a great extent mapped out by the use of echo-sounding and other equipment.

(2) Sampling the muds and animals of the seabottom

When only mud samples are needed from the seabottom, all that is to be done is to let down a metal tube of about 2-1/2 inches diameter; when such a tube is let down by force, it can easily core through a depth of about two feet (in soft mud), the mud getting in and filling the tube exactly in the same layers as found at the seabottom but the different layers may, to some extent get telescoped or pressed together; however, this telescoping of the layers may not matter for certain types of examination, and each layer can still give a picture of the soil condition as it prevails at the seabottom. There are various types of core samplers at present in use in different places, each offering some advantage over the other. In all these samplers the principle is the same as above, and for deeper regions of the sea there will be special devices for forcing the corer into the bottom.

One kind of piston core sampler has a piston which is pushed into the tube from the opposite end after the sampler is taken out of the sea so that the telescoped cylinder of bottom mud is forced out with its different depth regions held in tact; Such samples can be dried or preserved in formalin and kept indefinitely for later examination. Sometimes the coring tube itself is so made that it can be split into longitudinal halves when needed after taking out of the sea; the core of mud from inside the tube can thus be easily taken, and the different strata preserved for examination. When research expeditions make long sea voyages, the samples are always preserved, labelled and brought back to the research laboratory to be examined in detail or to be sent to interested specialists for the examination and reporting. The examination will include among other things, the nature and grain size of the mud, colour and chemical composition of the mud, and the animal and plant contents of the same in the different parts of the cored sample representing the different depths of the sampled sea bottom.

When there is no need to get stratified samples of the sea bottom, a mere grab type of sampler may be quite sufficient; grabs and dredges are common instruments used for getting samples of bottom animals from the muddy and sandy bottoms of the sea. A grab is somewhat like the implement used for removing silt at harbour entrances frequently, this being commonly called the "dredge". The naturalist's dredge is however a different kind of implement as will be shown below. The Danish Scientist Petersen was the first to use this instrument (i.e. the grab) with his own special design for taking collections of the bottom fauna from the sea. The special construction in this case is such that the instrument takes a sample of mud or sand with all its contents from an area of 0.1 sq. metre of the seabottom; there are two valves for the instrument and the entire set is sent down by means of a rope or chain in an open condition; there is an arrangement by which the valves are released as soon as the instrument touches the bottom; the two valves cut into the mud and move towards each other at the same time over a hinge by which they are connected; when the two valves meet each other they enclose between them all the mud and its contents found along their path as they tend to close themselves on each other; these valves normally dig to a depth of about 10 cm from the surface of the seabottom but if the mud is too hard the depth may be less, so that the amount of the mud finally enclosed between the two valves of the grab will depend on the consistency of the mud itself; in many normal areas of the sea-bottom the space between the two valves of grab will be full with the mud when the grab closes down after digging into the seabottom.

The grab samples are sieved for recovering the animal contents. Nests of sieves of graded dimensions can be used so that different sizes of the animals as well as the sand and mud particles may be sorted out for separate examination and further treatment. Dr. Petersen used the grab which has subsequently taken his name as "the Petersen Grab", to survey the quantitative distribution of bottom animals in the Danish waters in correlation with his plaice investigations; the plaice is an important bottom feeder and the density of the food organisms of the sea bottom is very important in the consideration of the fishery variations in that species. The Petersen's Grab can now be made in our country also; the firm of M/s. St. Vincent Industrials Ltd., at Calicut had manufactured a couple of these successfully as early as 1950 based on specifications supplied by the present writer.

An earlier instrument in use for sampling bottom animals is the *naturalists' dredge;* while the grab sampler is used for quantitative sampling, the dredge is useful for qualitative sampling; while the grab takes samples of small areas of only 0.1 sq. metre each time, the dredge can be dragged for any distance and the animals of the entire distance thus covered can be sampled. Again there are various models of dredges according to the purpose and the nature of the substratum; on gravel a completely metal netting can be used as the bag of the dredge; the dredge frame itself is a rectangular frame of flat iron which forms the mouth of the instrument as it is dragged along the seabottom. The long sides of the rectangular frame will have a sharp edge or toothed cutting edges. Tied to this frame is a bag of variable dimensions and various sizes of meshes. The frame is tied to a rope or chain and dragged along the seabottom; the mud and sand and the smaller animals are washed out of the dredge and whatever does not pass through the mesh is retained and hauled up, this being the sample as obtained to represent the area over which the dredge has been dragged. The sample is then washed and the animals are taken out and examined. The size of the dredge is variable according to the circumstances of its use and the main items that are to be sampled. Dredges of various kinds are used not only in research sampling but also in regular fishing, e.g. the oyster dredge used in oyster fishing. The Clams and chanks from a muddy bottom can also be fished by using a suitable dredge. The mouth of a dredge frame may vary from 6" x 2' to 1' x 5'. Larger dredges can be used only from mechanised vessels.

Like many other things instruments used for different purposes also undergo change and improvement from time to time. The dredging instruments also have had their historical "evolutionary" stages. The most modern of these different forms of dredge was designed and used by one Dr. Higgins of the United States of America who visited our country as a member of a team participating in the International Indian Ocean Expedition in the nineteen-sixties. The frame used by him was made of Stainless Steel and the instrument was so designed that it could sample a peculiar group of animals that live between the surface of the mud (at the bottom of the sea) and the water layer immediately covering it. These are animals of small size called the *Kinorhyncha* living at the interface between the mud and the water above it and looking somewhat like small worms. We have had before that, plankton sampling equipment and equipment for sampling the bottom fauna; but this peculiar interface group of animals could not at all be sampled by the methods existing earlier, until Dr. Higgins designed his most interesting net-cum-dredge sampler. Though we were taking dredge and grab samples for several years in the same areas of the sea, at Calicut, we had not even suspected the existence of these tiny animals in these areas! We had no idea about them and were losing them altogether because they were washed out while cleaning the grab samples, and in the case of the dredge samples they were completely left out by the sampling equipment.

Various kinds of molluscs, starfishes, brittlestars, and other animals of the benthos can be easily collected by means of a dredge of one kind or the other. But for capturing the nektonic animals that may be not at the surface but moving at various depths of water (and that do not always get into the usual plankton nets), special nets will have to be designed. The midwater trawl with variable meshes will serve this purpose; the beam trawl can be used for catching animals above the seabottom in the lower layers of the water but not those occurring in the benthos; this has a D-shaped frame with a bag-shaped net tied to it, and a beam on either side of the frame on the lower side, i.e. along the seafloor helps to keep the instrument above the mud without getting buried into it. The small beam trawl may have a length of 6-7 feet while at the upper limit, the large trawl may as long as 50-60 ft.

(3) Sampling the Plankton

There are special types of plankton nets for the capture of the planktonic organisms living in the water and floating at various levels in the sea. The main parts of a plankton net are the frame and the "bucket" or vessel which serves as the receptacle for the sample. The best plankton nets are made from good quality silk called the "Bolting Silk"; the net frame is made either of cane or of metal rod bent in the form of a ring of 1/2, 1, 2, 4 or more feet diameter. The net itself is a long cone-shaped bag attached to this ring at the wide end and it becomes narrow at the free end to which the bucket (which is a cylindrical vessel) is attached. The free end of the bucket is covered by a removable piece of bolting silk of variable mesh sizes. The frame of the net which is thus complete is tied up at three or four points by equal lengths of manila rope to a thicker main rope by which the net is operated and hauled when the dragging is over. As the plankton net is dragged or towed in the water, the organisms in the water are washed inwards towards the bucket and the water passes out through the netting. The towing of the net in the water is done by keeping it in the required position in the water after lowering it from the boat by means of the main connecting rope, and by moving the boat at a constant speed for a fixed period of time so that the sample is roughly quantitative for time as well as distance covered by the net during the dragging. Finally when the net is hauled up into the boat it is empty except for some organisms that may be sticking to the silk here and there, and the bucket itself will come up with a thick concentrated sample of planktonic organisms. The sizes of the organisms thus retained in the bucket depend on the mesh of the bolting silk used in the net and in the bucket at its end. Fine silk is used for phytoplankton material while a coarse mosquito netting may be enough if even the normal zoo-plankton is not needed but only larger forms are required such as the larvae of fishes; the contents of the plankton bucket are washed into a glass vessel and brought to the laboratory immediately for further examination. But it may often be necessary to fix or preserve the plankton sample in the sea itself so that the material may not deteriorate before reaching the laboratory. This difficulty is usually there if the sampling is done several miles away from the shore on a hot day and also when a number of samples are taken compelling stay at sea for a very long time. In the case of some expeditions the research ships may stay out at sea for several weeks or months at a time but in such cases facilities will usually be there for a preliminary examination of the material and of its sorting and preserving groupwise or otherwise so that on reaching the shore the samples so labelled and preserved can be sent to interested specialists for working out the same; the specialists may be from the same establishment or spread over an entire country or even over several countries. In all such cases where long preservation is required large quantities of formalin are taken on board the research vessels and using sea water for the dilution, 5% formalin is prepared and used for the preservation of the material. Depending on the nature of the organisms being preserved, the first lot of formalin used may have to be replaced after a week or two and fresh 5% seawater formalin added to replace the old formalin. Such renewal of formalin becomes very necessary when the first fixing-cum-preservation is done in a hurry and the organisms preserved are not fully cleaned before bottling, such things sometimes happening during rough sea and other such reasons beyond control, though only rarely.

Plankton nets can be used at all levels of the water in the sea, according to the workers' requirements. For strictly quantitative purposes vertical hauls are often taken by lowering the net to the seabottom and dragging the same slowly upwards trapping the organisms occurring in the column of water through which the net rises up. There are also what are known as the Vertical closing nets which are used for taking plankton samples from different layers of the seawater at the same time in a single haul of the net; these may be considered like a series of single nets made to sample at different depths and made so as to close separately as soon as the particular depths required are passed through; there are also nets which can be made to close at any stage of the operation and perhaps also single nets with many compartments, each compartment made to close at a particular depth after sampling upto that depth in the column of the seawater.

For ordinary plankton collection bolting silk of 60 meshes to the inch seems to be sufficient; a net for such purposes may have a frame of $1\frac{1}{2}$ feet diameter and five feet length from ring to bucket, the diameter of the ring gradually decreasing towards the bucket which is itself of a diameter of about 4 inches only.

Standardization of the mode of collection and in fact of all methods including those of preservation and examination will be necessary wherever we wish to compare the results of different regions with one another. Researches of the same kind will be going on in different parts of the world and it will be always helpful to compare the results of one place with another for drawing useful conclusions and modifying further plans of investigations so as to make the studies more purposeful and fully useful in an all-round sense. Such standardisation is usually achieved by liason with international organisations; the Indo-Pacific Fisheries Research Council and certain other organisations are helping in this direction. Japan has during the recent past designed a standardised plankton net called the "Moratoku" plankton net for international use; this has been accepted by the IPFC also.

A few decades ago Professor A.C. Hardy of the Hull University in England developed and used a special instrument called the *continuous plankton recorder* by means of which he was able to collect extensive samples from various parts of the North Sea covering complete distances of hundreds of miles covered by some commercial ships during the last world war. The instrument was such that it had only to be tied up on one side of the ship at the commencement of the voyage, kept working during the voyage and taken out at the end of the voyage; then a certain part of the inside of the instrument had to be taken out and sent to the scientist concerned or authorised to receive the same; this portion was something like a film spool of a photographic camera (but larger) and had to be replaced by a new one while starting on the next voyage. Once the finished spool reached the laboratory the researcher had only to spread out from this part a rolled ribbon of bolting silk which was to be unrolled inch by inch and examined under a special binocular microscope for the organisms on the ribbon. Each marked length of the ribbon would represent a certain distance traversed continuously in the sea, the exact distance being dependent upon the voyage and in all cases adjustable according to requirement. The roll of bolting silk would be already in formalin so that the material is well preserved and can be taken up for examination according to the convenience of the worker at his own time. A large number of voyages of the commercial ships between England and the continental countries on the other side of the North Sea were utilized like this and the very interesting results obtained are described by Professor Hardy and his colleagues in the earlier Bulletins of the Hull University on Marine ecology entitled *"Hull Bulletins of Marine Ecology"*. The present writer had the opportunity of visiting one of the Branches of the Hull University Marine Biological Laboratory, namely the Leith Branch at Edinborough and to witness a demonstration of the continuous Plankton recorder and an examination of the samples collected by the same; it was indeed a very educative visit, some time during 1947.

In all marine institutions some amount of plankton work may be said to be always going on in some form or other. Most of the marine animals spend at least a part of their larval lives in the plankton; and when we have to study the life histories of marine organisms, the nonnal source of material to us is the plankton from which the larvae are to be collected and studied, and sometimes reared also.

(4) Collections of water samples and reading temperatures

Taking samples of water from the surface is not a difficult problem; this can be done as from any tank or lake by dipping a suitable container in the water by hand and taking it out when it is filled; the container should of course be cleaned before use and rinsed with the seawater once or twice before the actual sample is taken. Water samples are normally taken in glass or plastic bottles. The rinsing with sea water at the sampling site assures that there is no source of contamination in the samples.

Special equipment is necessary for taking samples from deeper waters. The same holds good for taking temperatures of the sea water also. At the sea surface the temperatures are taken by the water in a bucket after rinsing the vessel a couple of times with the sea water of the same station; the thermometer is directly dipped in the sample of water immediately it is taken out and the temperature reading is read off and recorded.

There are what are called water sampling bottles for taking samples from deep waters, the important requirement here being that the sample should not have any chance of getting mixed up with any traces of the intermediate layers of the water during the passage of the bottle to the surface after sampling and retaining the bottom water within itself. Such bottles have also arrangements usually for reading the temperatures also built in within the water-bottle system. There may be occasions when only the sea water temperature is to be measured without necessarily taking out a sample of water also; in such cases what are known as deep sea reversing thermometers are the ones to be used; once released

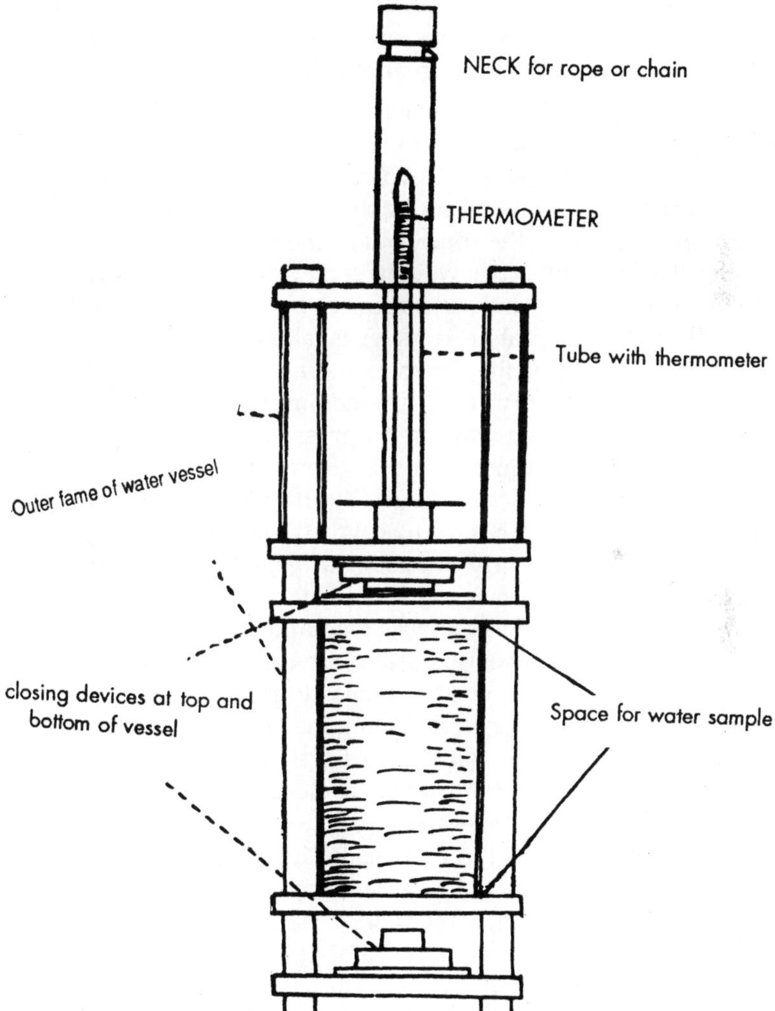

Fig. 58. Diagram of a Nansen-Pettersen Water Collection Bottle (In the Open Position).

in a certain position these reversing thermometers can be made to record the temperature at any depth and then made to reverse by sending a "messenger" (which is a ring-like metal piece of weight that can descend along the rope or chain so as to hit the apparatus below, this "hitting" forcing the equipment to reverse itself in position); this reversing keeps the temperature reading constant for recording when it is drawn up to the surface. Taking the temperatures of water samples seems often unnecessary these days as many of the larger research vessels in most countries are equipped with bathythermographic equipment so that continuous temperature values are available for all layers of the water in the investigated area; and such recordings are possible for as many stations are required without the need of too much manpower and motor power.

A water sampling bottle may be said to consist of a cylinder open at both ends, held firmly in a frame and with provision for closing the cylinder at any required level. As the bottle passes downwards through the column of water, different layers of the water will be passing through the cylinder as both the ends of this are kept open all along the route; but at the required level the cylinder is closed by means of a messenger (i.e. weight) sent down from above, and no fresh traces of water can enter the cylinder after this closure. The water bottle is hauled up in this condition and the contained water is taken out and analysed. Various models of such bottles exist in the international as well as the national market, though the latter source is yet rather too limited; improved models are frequently coming out and researches are always in progress on more and more perfection in these as in all other instruments and equipment. One of the well-known models is the *Nansen-Pettersen water sampler* , which is shown in the figure 58. In shallow waters even simpler bottles like those designed for oxygen sampling by Messrs Cassella of London early in this century can be used. This is a simple cylindrical vessel of metal having a screw-top with two openings on the lid, one raised on a prolonged tube about six inches long and the other flush with the top surface of the cylinder which is closed otherwise on all sides as well as the top and bottom; the top lid can be opened out so as to place a glass or plastic bottle inside the cylinder for the purpose of collecting the water sample into it; the flush opening on top of the lid of the cylinder has a short inward tube which leads inside the mouth of the glass bottle placed inside the cylinder; the top screw is now closed tight and a rubber tube is attached to the nozzle of the tube projecting out with the second opening of the cylinder at its end; the entire apparatus is now attached to a rope or chain through a handle provided on top of the cylinder independent of the two openings mentioned above. There is a small twine attached to the rubber tube attached to the projecting nozzle of the upper side of the casella bottle and after sending down the bottle to the required level this twine is pulled out and after the time for the bottle to be filed, the entire bottle is dragged out of the water as quickly as possible. The lid is unscrewed and the water in the glass or plastic bottle inside the cylinder is collected and analysed. The principle here is that because of the two openings of the upper part of the apparatus being at two different levels water will pass inside through one of these and comes out from the other opening; the inward current will fill the bottle placed inside while the projecting nozzle will bring out air and water pushed out due to difference of pressure. The mechanism is arranged in such a way that the water currents are set in motion only after the bottle reaches the bottom; then the bottle is dragged up so quickly that in shallow waters there is not much contamination during the passage of the bottle in its upward journey to the surface from the bottom. There is of course at least theoretically scope for some trace of contamination with column water if this equipment is used in deep waters, but in shallow depths there is not much scope for such error. The sheeting used for the cylinder must be thick so as to give a reasonably heavy weight to it for sinking easily.

Like all samples from the sea, water is also better analysed in the fresh condition as soon as it is taken out from the sea. Certain items like temperature, hydrogen-ion-concentration and oxygen are to be determined immediately. For items like salinity, total phosphorus etc., samples may be preserved in a prescribed manner subject to certain precautions. Usually the elementary hydrological work will include the determination of temperature, specific gravity, p^H, salinity, oxygen etc., any number of other determinations being added according to requirement and facilities available. These may include for instance, phosphate, silicate, nitrite, nitrate, carbon-di-oxide, hydrogen sulphide, etc. The hydrogen-ion concentration is well determined these days by using one of the electrical p^H meters; in the absence of these, ordinary colour comparators with BDH standards or even P^H-paper can be used. Salinity of sea water is usually determined by titrating 10ml of the water against silver nitrate using Potassium dichromate as indicator; the end point of the titration gives the chlorinity values by equation and from this the salinity can be calculated.

Various text books and other publications are in the market describing methods of seawater analysis.

Bibliography

Ali, M.A. (1956 a): Additions to the sponge fauna of Madras. *J. Madras Univ., B, 26,* No (2), pp. 289-301.

_____ (1956b): Development of the monaxonid sponge *Lissodeudoryx similis* (Thiele). *J. Madras Univ., B, 26,* No. (3), pp. 553-581.

_____ (1960): Influence of environment on the distribution and form of sponges. *Nature,* London. 186, No. (4719), pp. 177-178.

Allen E.J. (1928): *Science of the sea.* Clarendon Press, Oxford, xxiii + 502 pp.

Alagarswami, K. and S.Z. Qasim (1973): Pearl culture—its potential and implications in India. *Indian J. Fish., 20,* (2), pp. 533-550.

Annandale, N. (1915): Fauna of the Chilka lake. Sponges. *Mem. Indian. Mus., 5,* pp. 21-54.

Bal, D.V. and K.V. Rao (1984): *Marine Fisheries.* Tata McGraw-Hill Publishing Company Limited, New Delhi. xii + 470 pp.

N. Balakrishnan Nair and D.M. Thampy (1980): *A textbook of Marine Ecology.* The Macmillan Company of India Ltd (Printed at Prabhat Press, Meerut), xvi + 352 pp.

Bapat, S.V. (1970): The Bombay duck *Harpodon nehereus* (Ham). *Bull. Cent. Mar. Fish. Res. Inst., 21,* pp. 1-66.

Bhimachar, B.S. and P.C. George (1950): Abrupt setbacks in the fisheries of the Malabar and Kanara coasts and "red water" phenomenon as their probable cause. *Proc. Ind. Acad. Sci. B, 31,* pp. 339-350.

Bhimachar, B.S. and G. Venkataraman (1952): A preliminary study of the fish populations along the Malabar coast; *Proc. Natn. Inst. Sci. India, 18 B,*(6), pp. 627-655.

Burton, M (1937): Supplement to the fauna of the Krusadai Island in the Gulf of Mannar—Porifera. *Bull. Madras Govt. Mus.,* N.S., *1,* (2), pt. 4, 58 pp + 9 plates.

Burton, M. (1959): Sponges. *John Murray Expedition 1933-34. Scientific Reports. Brit. Mus. Nat. Hist., 10,* (5), pp. 151-281.

Burton, M. and H.S. Rao (1932): Report on the shallow water marine sponges in the collection of the Indian Museum 1. *Rec. Ind. Mus., 34,* pp. 299-356.

Colin Nicol J.A. (1967): *The biology of marine animals*, Pitman, London. xi + 699 pp.

Chari, S.T. (1948): Nutritive value of some west coast food fishes of the Madras Province. *Ind. J. Med. Res., 36*, pp. 253-259.

Carson, Rachel (1959): *The edge of the Sea*. New American Library Mentor Edition, 238 pp. (First published 1955).

Day, Francis (1958 reprint): *The Fishes of India*. in 2 volumes and a supplement. Vol. 1, xx + 778 pp; vol. II, 195 plates; supplement 779-816 pp. Vols. I & II published by William Dawson & Sons Ltd., London; Supplt. issued as photo-offset reprint by the Indian Society of Ichthylogists, Madras (1980).

Das, S.M. (1936): *Herdmania* (a monascidian of the Indian Seas). *Indian Zool. Mem., 5*, pp. 1-103.

Dhulkhed, M.H. and C.N. Nagesh (1976): Serological studies on oil sardine (*Sardinella longiceps*)and mackerel (*Rastrelliger kanagurta*). (*Fish. Technol., 13*, (1), pp. 9-12.

Dhulkhed, M.H. and S. Narasimha Rao (1976): Electrophoretic studies on serum proteins of Oil sardine (*Sardinella longiceps*) and mackerel (*Rastrelliger kanagurta*). *Fish. Technol., 13*, (1), pp. 16-19.

Fauvel, P. (1930): Annelida Polychaeta of the Madras Government Museum. *Bull. Madras Govt. Mus., N.S. 1*, (2): pp.1-72.

Fauvel, P. (1932): Annelida Polychaeta of the Indian Museum, Calcutta. *Mem. Ind. i an. Mus., 12*, pp. 1-262.

Fauvel, P. (1953): *Fauna of India. Annelida Polychaeta*. Indian Press Ltd., Allahabad, pp. 1-507.

Gardiner, J. S. (1903-1906): The fauna and geography of the Maldive and Laccadive archipelagoes. vol. I, Cambridge University Press, Cambridge.

George, P.C. (1953): The marine plankton of the coastal waters of Calicut with observations on the hydrological conditions. *J. Zool. Soc. India, 5*, pp. 76-107.

Gnanamuthu, C.P. (1943): The foraminifera of the Krusadai Island (in the Gulf of Mannar). *Bull. Madras Govt. Mus., N.S., 1*, (2). pt. 5, pp 21 + 4 plates.

Gnanamuttu, J.C. (1972): On the occurrence of *Rastrelliger faughni* Matsui in the Indian Waters. *Indian J. Fish., 18*, (1 & 2), (1971), pp. 170-173.

Gopala Aiyar, R. (1966): *Salmacis (The Indian Sea Urchin)*, With frontispiece and 47 text figures. Revised by P.N. Ganapati, 72 pp. Zoological Society of India, Indian Zoological Memoirs No. 7. Calcutta.

Gopinadha Pillai, C.S. (1969): Corals and coral reefs. *Proc. Symp. Corals and coral reefs*. Mar. Biol. Association of India, Mandapam Camp. pp. 9-17.

Gopinadha Pillai, C.S. (1972): Stony corals of the seas around India. *Proc. Symp. Corals and Coral reefs*. Mar. Biol. Association of India, Mandapam Camp. pp. 191-216.

Gopinadha Pillai, C.S. (1973): Coral resources of India with special reference to Palk bay and the Gulf of Mannar. *Proc. Symp. Living Resources of the seas around India*. Special Publication, CMFRI (ICAR), pp. 700-705.

Gravely, F.H. (1941): Shells and other animal remains of the Madras beach I. Groups other than snails etc. *Bull. Madras Govt. Mus.*, N.S. *5*, (1), 112 pp.

Gravely, F.H. (1942): Shells and other animal remains of the Madras beach. II. Snails etc. (Mollusca Gastropoda). *Bull. Madras Govt. Mus.*, N.S. *5*, (2), 110 pp.

Hartman, Olga (1974): Polychaetous annelids of the Indian ocean including an account of species collected by members of the International Indian Ocean Expedition 1963-64 and a catalogue and bibliography of the species from India. *J. mar. biol.Ass. India, 16*, (1), pp. 191-252 (publ. 1975).

Hartman, Olga (1974 b): Polychaetous annelids of the Indian ocean including an account of species collected by members of the International Indian Ocean Expedition 1963-64 and a catalogue and bibliography of the species from India. Part II. *J. mar. biol. Ass. India, 16*, (2), pp. 609-644 (published 1976).

Hickson, S.J. (1931): Three species of Alcynoaria from the Gulf of Mannar. *Bull. Madras Govt. Mus., 1*, (2), pt. 2, pp. 81-89.

Hefford, A.E. (1943): Report on the work of the Steam Trawler "William Carrick", May 1921 to February 1922 (Govt. Press, Bombay, 1923) (reprint).

Herdman, W.A. (1903-1906): *Rep. Govt. Ceylon Pearl Oyster Fish. Gulf of Mannar, 1, 2, 3 & 4.*

Herdman, W.A. and Hornell, J. (1906): Pearl Production. *Rep. Govt. Ceylon Pearl Fish. Gulf of Mannar, 5*, pp. 1-42.

Hornell, J. (1915): The recent pearl fishery in Palk bay with biological notes on the pearl oyster. *J. Asiatic. Soc. Bengal, N.S. 11*, pp. 153-164.

Hornell, J. (1916): Professor Huxley and the Ceylon pearl fishery with a note on the forced or cultural production of free spherical pearls. *Madras Fish. Bull. 8*, pp. 93-104.

Hornell, J. (1922): The Indian pearl fisheries of the Gulf of Mannar and Palk bay. *Madras Fish. Bull. 16*, pp. 1-188.

Jacob, T., K. Alagaraja and K.N.Kurup (1983): Marine Fishery statistics in India - Present status. *Mar. Fish. Infor. Serv. T. & E. Ser. 46*, pp. 6-11.

James, P.S.B.R. (1967): *The ribbon fishes of the family Trichiuridae.* Memoir I of the Marine Biological Association of India, Mandapam Camp. 228 pp.

Jayaraman, R. and G. Seshappa (1957): Phosphorus cycle in the sea with particular reference to tropical inshore waters. *Proc. Indian Acad. Sci. 46,B.* pp. 110-125.

Jhingran, V.G. (1983): *Fish and Fisheries of India* (Revised and enlarged second edition). 666 pp. Hindustan Publishing Corporation, Delhi.

John, C.C. (1933): *Sagitta* of the Madras coast. *Bull. Madras Govt. Mus., N.S. 3*, (4), pp. 1-10.

Jones, S. and S.K. Banerji (1973): A review of the living resources of the Central Indian Ocean. *Symp. Liv. Res. Seas around India,* CMFRI, Cochin (Special Publication), 1973, pp. 2-5.

Jones, S. (1986): Lakshadweep—General features and some considerations. *Mar. Fish. Infor. Serv. T & E. Ser.* No. *68* (CMFRI).

Jones, S. and M. Kumaran (1980): *The fishes of the Laccadive Archipelago.* Nature, conservation and Aquatic Sciences Service, Trivandrum 760 pp.

Kameswara Rao, K. (1969): Foraminifera of the Gulf of Cambay. *J. Bombay Nat. Hist. Soc. 66,* (3) , Dec. 1969, pp. 584-596.

Kameswara Rao, K. (1970): Foraminifera of the Gulf of Cambay. *J. Bombay Nat. Hist. Soc. 67,* (2), Aug. 1970, pp. 259-273.

Kameswara Rao, . (1972): Planktonic foraminifera in sediment samples from the eastern Arabian sea. *Indian J. Mar. Sci., 1,* (June 1972), pp. 1-7.

Kewalramani, H.G., P.V. Wagh and M.R. Ranade (1960): Taxonomy of the lugworm found off Bombay. *J. Zool. Soc. India, 11,* (2), pp. 109-116.

Krishna Menon, M. (1933): The life histories of the Decapod Crustacea from Madras. *Bull. Madras Govt. Mus., N.S. 3,* (1), 28 pp.

Krishna Menon, M. (1937): Decapod larvae from the Madras Plankton. *Bull. Madras Govt. Mus., N.S., 3,* (5), 55 pp.

Krishna Menon, M. (1940): Decapod larve from the Madras Plankton, II. *Bull. Madras Govt. Mus. N.S., 3,* (6), 47 pp.

Krishna Moorthy, B. (1962): Salinity tolerence and volume regulation in four species of polychaetes. *Proc. Ind. Acad. Sci. B, 56,* (6), pp. 363-371.

Krishna Moorthy, B. (1963): On the distribution of six species of polychaetes in the Adyar estuary. *J. mar. biol. Ass. India, 5,* (1), pp. 97-102.

Krishna Moorthy, B. (1963): Chloride regulation in *Marphysa gravelyi* Southern (Polychaeta). *Curr. Sci., 35,* (3), pp. 72-73.

Lal Mohan, R.S. (1984): Sciaenidae. *FAO species identification sheets for fishery purpose.* Western Indian Ocean (Fishing area 51). FAO, Rome, 77 pp.

Lal Mohan, R.S. (1982): An illustrated synopsis of the fishes of the family Sciaenidae of India. *Indian J. Fish., 28* (1 & 2), pp. 1-24.

Lal Mohan, R.S. (1980): Some observations on the sea cow *Dugong dugon* from the Gulf of Mannar and Palk bay during 1971-1975. *J. mar. biol. Ass. India, 18,* (2), (1976), pp. 391-396.

Lal Mohan, R.S. (1972): A synopsis of the Indian genera of the fishes of the family Sciaenidae. *Indian J. Fish., 16,* (1 & 2), (1969), pp. 82-98.

Lal Mohan, R.S., D.B. James and K.K. Appukuttan (1969): Bibliography of the Indian Ocean. *Bull. Cent. Mar. Fish. Res. Inst., 11,* 1-176.

Mahadevan, S. and K.N. Nayar (1973): Pearl oyster resources of India. *Proc. Symp. Liv. Res. Seas around India.,* pp. 659-691. (Spl. Publication of the CMFR Institute, Cochin).

Menon, M.G.K.(1930): Scyphomedusae of the Krusadai Island. *Bull. Madras Govt. Mus. N.S., 1,* (2), pt. 3, pp. 9.

Menon, M.G.K. (1932): The Hydromedusae of Madras. *Bull. Madras Govt. Mus., N.S., 3,* (2), 32 pp.

Misra, K.S. (1962): An aid to the identification of the common commercial fishes of India and Pakistan. *Rec. Indian Mus., 57,* pp. 1-320 (1959)

Mukundan, C. (1967): Plankton of Calicut inshore waters and its relationship with coastal pelagic fisheries. *Indian J. Fish., 14,* (1 & 2), pp. 271-292.

Nair, R.V. (1973): *Indian Sardines.* (Biology and Fishery). *CSIR Zoological Monograph* no. 2, CSIR, N. Delhi., 107 pp.

Nair, R.V. *et al.* (1970): The Indian Mackerel. *Bull. No. 24,* CMFRI, Cochin, 102 pp. (Mimeo).

Nair, R.V. and R.S. Lal Mohan (1977): Studies on the vocalisation of the sea cow *Dugong dugon* in captivity. *Indian J. Fish., 22,* **(1 & 2) (1975)**, pp. 277-278.

Narasimha Rao, S. and M.H. Dhulkhed (1976): Electrophoretic characteristics of oil sardine (*Sardinella longiceps*) and mackerel (*Rastrelliger kanagurta*) eye lens proteins. *Fish. Technol., 13,* pp. 13-15.

Noble, A. (1985): The mackerel fishery—a short review. *Mar. Fish. Infor. Serv. T & E. Ser. No. 63,* May - June 1985, pp. 1-6 (CMFRI).

Norman, J.R. (1928): The flatfishes (Heterosomata) of India with a list of the specimens in the Indian Museum, Pt. 2, *Rec. Indian Mus., 30,* pp. 173-216.

Prasad, R.R. (1956): Further studies on the plankton of the inshore waters off Mandapam. *Indian J. Fish.,* 3, (1), pp. 1-42.

Prasad, R.R. (1957): Seasonal variations in the surface temperatures of the seawater at Mandapam from January 1950 to December 1954. *Indian J. Fish., 4,* (1), pp. 20-31.

Prasad, R.R. and P.V.Ramachandran Nair (1974): India and the Indian Ocean fisheries. *J. mar. Biol. Ass. India, 15,* (1), pp. 1-17.

Qasim, S. Z., (1977): Biological productivity of the Indian Ocean. *Indian J. Mar. Sci., 6,* (2) : pp. 122-137.

Ramachandran Nair, P.V. (1970): Primary Productivity in the Indian Seas. *CMFR Institute Bulletin* No. 22, Cochin.

Ramachandran Nair, P.V., C.V.Gopinathan, and V.K. Balachandran (1978): Studies on phytoplankton productivity and the estimation of potential resources. *CMFR Institute Spl. Publication No. 3,* pp. 109-116.

Ramachandran Nair, P.V., Sydney Samuel, K.J. Joseph and V.K. Balachandran (1973): Primary production and potential fishery resources in the seas around India. *Proc. Symp. Living Resources of the seas around India. CMFR Institute Spl. Publication* pp. 184-198.

Ramachandran Nair, P.V. and V.K. Pillai (1983, publd. 1987): Productivity of the Indian seas. *J. mar. biol. Ass. India, 25,* (1 & 2), pp. 41-50.

Ranade, M.R. (1952): Occurrence of *Arenicola* in Bombay. *Curr. Sci., 21,* p. 165.

Seshappa G. (1966): Raciation and speciation in the littoral region. *Proc. All-India Congress Zool.* Varanasi, 1962. Pt. 2.

Seshappa, G. (1976). Reproduction in *Littorina saxatilis rudissima* (Bean) and two other subspecies of *Littorina saxatilis* (Olivi) at Cullercoats, England. *Indian J. Mar. Sci., 5,* Dec. 1976., pp. 222-229.

Seshappa, G. (1953): Observations on the physical and biological features of the inshore seabottom along the Malabar coast. *Proc. Natn. Inst. Sci. India, 19,* pp. 288-301.

Seshappa, G. and R. Jayaraman (1956): Observations on the composition of bottom muds in relation to the phosphate cycle in the inshore waters of the Malabar coast. *Proc. Indian Acad. Sci., 43, B,* (6), pp. 288-301.

Seshappa, (1970): Mackerel, an important fish in the seas around India. *Indian Farming, 20,* (1), pp. 39-41.

Seshappa, G. (1972): The problem of age determination in the Indian mackerel, *Rastrelliger kanagurta* by means of scales and otoliths. *Indian J. Fish., 16,* (1 & 2) pp. 14-28 (1969).

Seshappa, G. (1973): The flatfish resources of the seas around India. *Proc. Symp. Living Resources of the Seas around India.* Spl. Publn., CMFRI, Cochin, pp. 470-482.

Seshappa, G. (1985): On the homogeneity of the mackerel populations at Calicut during the years 1969 to 1976 as determined on the basis of C/L, C/W and TL/SL ratios. *Indian J. Fish. 32,* (3), pp. 359-374.

Seshappa, G. and B.S. Bhimachar (1951): Age determination studies in fishes by means of scales with special reference to the Malabar sole, *Cynoglossus semifasciatus* Day. *Curr. Sci., 20,* pp. 260-262.

Seshappa, G. and B.S. Bhimachar (1953): Studies on the age and growth of the Malabar sole, *Cynoglossus semifasciatus* Day. *Indian J. Fish., 1,* (1), pp. 145-162.

Seshappa, G. and B.S. Bhimachar (1955): Studies on the fishery and biology of the Malabar sole, *Cynoglossus semifasciatus Day Indian J. Fish., 2,* pp. 180-230.

Selvanathan, M. (1976): The occurrence of *Arenicola bombayensis* Kewalramani et al. (Family Arenicolidae, Polychaeta) at Muttam in southwest India. *J. Bombay Nat. Hist. Soc., 73,* (1), pp. 94-97.

Sewell, R.B.S. (1932 etc.): The John Murray Expedition to the Arabian Sea. In *The Scientific Reports of the John Murray Expedition 1933-34.* British Museum of Natural History, London, Vols. 1-10.

Sewell, R.B.S. (1929): Geographic and oceanographic research in Indian waters. V. The temperature and salinity in the surface waters of the Bay of Bengal and the Andaman sea with reference to the Laccadive sea. *Mem. Asiat. Soc. Beng., 9,* pp. 207-355.

Silas, E.G. (1982): Proceedings of the workshop on acquisition and dissemination of data on marine living resources of Indian seas. *Mar. Fish. Infor. Serv. T & E. Ser. No. 46.*

Silas, E.G. (1983): Sea turtles of India—need for a crash programme on conservation and effective management of the resource. *Mar. Fish. Infor. Serv. T & E Ser. No. 50,* June, 1983.

Silas, E.G. (1983: (Ed.) Proceedings of the workshop on acquisition and dissemination of data on marine living resources of the seas: *Mar. Fish. Inf. Serv. T. & E. Ser.* No. 46, 1982, 49 pp.

Silas, E.G. (1985): Tuna fisheries of the exclusive economic zone of India: Biology and stock assessment. *Bull. No. 36,* CMFRI, Cochin.

Silas, E.G. and P.P. Pillai (1982):Resources of tunas and related species, and their fisheries in the Indian ocean. *Bull. Cent. Mar. Fish. Res. Inst., 32,* pp. 1-174

Silas, E.G., Daniel Selvaraj and A. Reghunathan (1969): Rare chimaeroid and Elasmobranch fishes from the continental slope off the west coast of India. *Curr. Sci., 5,* March 5, 1969, pp. 105-106.

Smith, J.L.B. (1965): *The seafishes of southern Africa.* 5th edn. Central News Agency Ltd., S. Africa, 589 pp; 4th edn. 1961, 580 pp.

Smith, J.L.B. (1955): Live Coelacanths. *Nature,* London, 176, p. 473.

Subrahmanyan, R. (1946): A systematic account of the marine plankton diatoms of the Madras coast. *Proc. Indian Acad. Sci. 24, B,* 85-96.

Subrahmanyan, R. (1959): Studies on the phytoplankton of the west coast of India, pts. I & II. *Proc. Indian Acad. Sci., 50, B,* pp. 113-187 & 189-252.

Subrahmanyan, R. and A.H. Viswanatha Sarma (1960): Studies on the phytoplankton of the west coast of India Pt. III Seasonal variation of the phytoplankton and the environmental factors. *Indian J. Fish., 7,* pp. 307-336.

Subrahmanyan, R. and A.H. Vishwanatha Sarma (1967): Studies on the phytoplankton of the West coast of India Pt. IV Magnitude of the standing crop for 1955-62 with observations on nannoplankton and its significance to fisheries. *J. mar. biol. Ass. India., 7,* (2), pp. 406-419.

Swaminathan, M.S. (1983): *Science and the Conquest of Hunger.* xvi + 508pp. (Concept Publishing Company, New Delhi).

Tampi, P.R.S. (1946): *Studies on polychaetes.* M.Sc. Thesis, University of Madras.

Tampi, P.R.S. (1963): On the swarming of heteronereids in the Gulf of Mannar. *J. mar. biol. Ass. India, 5,* (2), pp. 246-250.

Tampi, P.R.S. and K. Rengarajan (1964): Some polychaetous annelids from Andaman waters. *J. mar. biol. Ass. India, 6,* (1), pp. 98-121.

Thomas, P.A. (1968): The sponge resources of India. *Abstract. Symp. Liv. Res. Seas around India,* CMFRI, Cochin.

Thomas, P.A. (1970a): Studies on Indian sponges. I. Two new species of silicious sponges belonging to the genera *Echinodictyum* Ridley and *Rhabderemia* Topsent (Class Domospongiae), *J. mar. biol. ass. India,* 10, (2), pp. 245-249.

_____ (1970b) : Studies on Indian sponges II. Two new species of silicious sponges belonging to the genera *Aka* de Laubenfels and *Damiorina* Burton. *Ibid., 10,* (2), pp. 250-254.

Thomas, P.A. (1970 c): Studies on Indian sponges III. Two silicious sponges of the family Ophlitaspongillidae de Laubenfels (Class Demospongiae Sollus, Order: Poecilosclerida Topsent), *Ibid.*, *10*, (2), pp. 255-259.

Thomas, P.A. (1970d): Studies on Indian sponges IV. Additions to the genus *Corticium* Schmidt. Ibid., *10*, (2), pp. 260-263.

Thomas, P.A. (1970e): Studies on Indian sponges V. Two new records of silicious sponges belonging to the families Myxillidae Heutschel and Spirastrellidae Heutschel from the Indian region. *Ibid.*, *10*, (2), pp. 264-268.

Thomas, P.A. (1976): The history of spongology in the Indian Ocean. *J. mar. biol. ass. India*, *18*, (3), pp. 610-625.

Thomas, P.A. (1985): Demospongiae of Minicoy Island (Indian Ocean) part I orders Keratosida and Haplosclerida. *J. mar. biol. ass. India*, *21*, (1 & 2), pp. 10-16.

Thomas, P.A. and Rani Mary George (1987): *Mar. Fish. Inf. Serv. T& E. Ser. No. 74*, pp. 1-13, CMFRI, Cochin.

Various Authors (R.V. Nair, K.V. Rao, V. Balakrishnan, G. Venkataraman, S.K. Banerji and K.V. Narayana Rao). (1970): *The Indian Mackerel*. Bull. No. 24, CMFRI, Cochin, 102 pp.

Index